TREE THIEVES

CRIME AND SURVIVAL IN NORTH AMERICA'S WOODS

LYNDSIE BOURGON

Little, Brown Spark
New York Boston London

Little, Brown Spark
Hachette Book Group
1290 Avenue of the Americas, New York, NY 10104
littlebrownspark.com

First Edition: June 2022

Little, Brown Spark is an imprint of Little, Brown and Company, a division of Hachette Book Group, Inc. The Little, Brown Spark name and logo are trademarks of Hachette Book Group, Inc.

The publisher is not responsible for websites (or their content) that are not owned by the publisher.

The Hachette Speakers Bureau provides a wide range of authors for speaking events. To find out more, go to hachettespeakersbureau.com or call (866) 376-6591.

Map by Jeffrey L. Ward

ISBN 9780316497442
LCCN 2021951367

Printing 1, 2022

LSC-C

Printed in the United States of America

For my parents, who prepared me well
for the journey

"We have mixed our labour with the earth, our forces with its forces too deeply to be able to draw back and separate either out."

—Raymond Williams,
Culture and Materialism

CONTENTS

CONTENTS

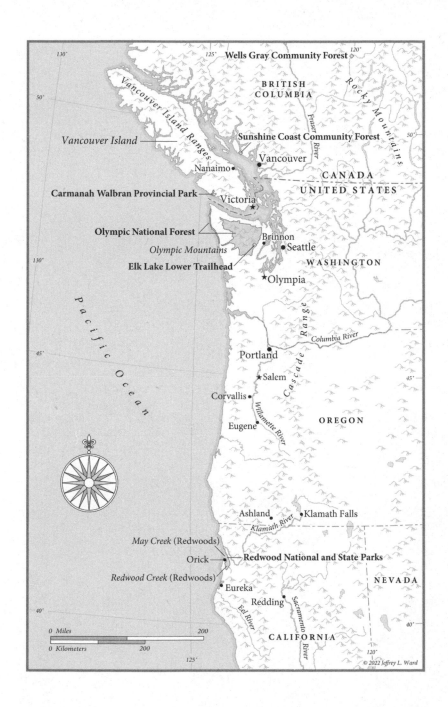

CHARACTERS

IN THE PAST

NEWTON B. DRURY: Executive director, Save the Redwoods
League; fourth director of the US National Park Service
ENOCH PERCIVAL FRENCH: First ranger-supervisor of
Northern California Redwood State Parks
MADISON GRANT: Cofounder, Save the Redwoods League
JOHN C. MERRIAM: Cofounder, Save the Redwoods League
HENRY FAIRFIELD OSBORN: Cofounder, Save the Redwoods
League
EDGAR WAYBURN: President of the Sierra Club (1961–1964)

IN THE FOREST

EMILY CHRISTIAN: Redwood National and State Parks
(RNSP) ranger
TERRY COOK: Danny Garcia's uncle
LAURA DENNY: Former RNSP ranger

DANNY GARCIA: A former "Outlaw"

CHRIS GUFFIE: Also known as "the Redwood Bandit"

JOHN GUFFIE: Chris Guffie's father

DEREK HUGHES: Another former "Outlaw"

BRANDEN PERO: Former RNSP ranger

PRESTON TAYLOR: Bear researcher, Humboldt State University

STEPHEN TROY: Chief ranger, RNSP

ROSIE WHITE: Former RNSP ranger

IN HUMBOLDT COUNTY

JUDI BARI: Earth First! activist

RON BARLOW: Lifelong Orickite; rancher

DARRYL CHERNEY: Earth First! activist

STEVE FRICK: Former logger

CHERISH GUFFIE: Terry Cook's girlfriend; Chris Guffie's ex-wife

JIM AND JUDY HAGOOD: Owners of Hagood's Hardware

JOE AND DONNA HUFFORD: Longtime Orick residents

LYNNE NETZ: Derek Hughes's mother

TREE
THIEVES

Prologue

MAY CREEK

At night, the treacherous curves of northern California's Redwood Highway unfurl before the probing reach of headlights. With little warning as to what's ahead, it's easy to miss turnoffs, so a small truck navigates the highway slowly, inching toward May Creek through the pitch-darkness of a damp winter night in 2018.

Just after midnight, the truck turns onto a lush wayside. It tips slightly as the driver pulls along the left-hand side of a metal gate, the tires toppling a small pile of rocks. The ground is soft enough that the tire grooves leave a lasting imprint in the earth. The driver points the truck back toward the road. Then it's dark again.

A narrow clearing stretches for about 100 yards—an old, decommissioned logging road that's been left to rewild and grow over. Climbing down from the truck, the driver finds a short trail beneath his feet, each side of the path lined with sword fern and clover, wallpapered in layers of redwood bark, though none of that is visible in the darkness. The floor is so thickly carpeted with foliage that his steps are muffled as he walks forward.

The man is lanky, his hair buzzed short, and he wears a

sweatshirt. He stands in the dark clearing, waiting for the truck's passenger to join him. The only light shines from headlamps.

Both men start to climb a nearby hill, one toting a chain saw. They walk through a thick tangle of branches and forest-floor debris, arms brushing up against red alder and vine maple. They are not going far, only about 75 yards, heading east and uphill from the highway and clearing. There is no official trail here, no campgrounds nearby; any stars that might peek through the thick Pacific fog are hidden by a thick treetop canopy.

They stop at the foot of a large, ancient redwood stump. One fires up the chain saw and the high-pitched buzz of the engine echoes loud across the clearing. No one driving along the Redwood Highway would be able to hear the strained noises of metal teeth biting into the deep ocher wood of the tree's trunk.

The trunk is about 30 feet in diameter and rooted at the edge of the hill. The man with the chain saw takes a short step down and leans into the incline. He begins to slice the base of the trunk vertically, on the side that faces away from the faint foot-path. His work is meticulous and neat: he carves squares with straight edges. Slowly the trunk is cleaved into fragments, falling to the forest floor like a glacier calves bergs into water. The logger's companion stands guard, and throughout the night the pair barely talk. Eventually they amass a pile of heavy rectangular blocks, some of which they push down toward the truck, slowly flipping the sections as they flop down the hill. They load the wood into the truck bed and drive away.

Back in the woods, the centuries-old redwood trunk remains with a third of its body poached: a gaping wound.

PART I

ROOTS

Chapter 1

CLEARANCES

The first case of tree theft I ever encountered occurred within the stands of ancient old-growth on the southwest shores of Vancouver Island, in Ditidaht territory. One day in the spring of 2011, a hiker in British Columbia's Carmanah Walbran Provincial Park noticed the smell of fresh sawdust in the air, and as he walked he spotted felling wedges—tools used to guide a tree's fall in a particular direction—thrust into the body of an 800-year-old red cedar. With the right wind, the tree, rising about 160 feet tall, could easily tip over. The wedges had shifted the tree from towering sentinel in lush rainforest to teetering public danger. BC Parks rangers were forced to down the cedar themselves. They left the tree on the forest floor to decompose, recycling back into the earth over the next hundred years.

It wouldn't last anywhere near that long: just 12 months later, most of the trunk was gone. After the tree was felled, poachers entered the park and sawed the trunk (or "bucked" it) into portable pieces, leaving a trail of sawdust and abandoned equipment behind. Ironically, by honoring their mandate of safety and conservation, BC Parks had made it easier for the tree to be stolen.

A local environmental group, the Wilderness Committee, sounded a public alarm about the poaching, and a press release sent out to journalists landed in my in-box. A decade later, no one has been charged under British Columbia's Forest and Range Practices Act with the crimes that took place in the Carmanah Walbran that night: unauthorized timber harvest from public property, and vandalizing timber. The cedar is long gone—sold to a local sawmill in the dead of night, or to an artisan who kept it in their shop, or turned it into shingles, or a clock, or a table.

Since then, I have watched a spate of wood poaching sweep North America: in the Pacific Northwest, the lush forests of Alaska, and the timber stands of the eastern and southern United States. Timber poaching happens everywhere, on vastly different scales, throughout the seasons—one tree taken here, another there. It has become "a problem in every national forest," according to forest officials, and it runs the gamut from the seemingly minute—cutting down a small Christmas tree in a park near your city, for example—to the large-scale devastation of entire groves.

In North America, the scale of timber poaching varies by region: In eastern Missouri, timber theft has become a frequent problem in Mark Twain National Forest, where in 2021 a man was charged with cutting down 27 walnut and white oak trees inside the park over the course of six months, then selling them to local mills. In New England, the primary victims are cherry trees. In Kentucky, the bark is stripped off the slippery elm tree for use in herbal remedies and diet supplements. Bonsai have disappeared from a museum garden in Seattle, palm trees from Los Angeles yards, a rare pine from an arboretum in Wisconsin, ancient alligator junipers from Prescott National Forest in Arizona. In Hawaii, koa trees—prized for their fine-grained red wood—are stolen from the rain forest. In Ohio, Nebraska, Indiana, and

Tennessee, I found the stumps of black walnut and white oak. None of these trees were rooted in logging land—all had been afforded some measure of protection, meaning they mattered to someone and some place.

Deep in the woods, there is other natural theft, too. Moss is sold to florists for about $1 per pound; in one case a poacher was caught with 3,000 pounds in the bed of his pickup truck. Across the southeastern United States, poachers rake up and sell the needles from longleaf pine, a resource dubbed "brown gold." Boughs off tree limbs, mushrooms, grasses, ferns—all are illegally traded forest products. Sometimes the very tops of spruce or fir trees are lopped off and sold as Christmas trees, or the tips of branches removed and turned into potpourri.

Forests are managed on stratified bureaucratic levels that, at points, overlap and collaborate. There are private property owners and forests managed by logging firms. There are also regional forests that fall under the jurisdiction of municipalities, states, or provinces. Then there's the National Park Service, the Forest Service, the National Monuments; in Canada there are Crown lands, national parks, and nature reserves. In the United States, most forests are privately owned and managed as forests or timber land. But in the western half of the country, most forest land is held by the federal and state governments— 70 percent of the forests there are publicly owned, compared with just 17 percent in the East.

It's easiest to understand these protective layers by considering the larger entity each organization falls under. For instance, the Forest Service is nestled within the Department of Agriculture. As such, the trees on Forest Service land are managed like a crop—a product that is grown and harvested and consumed. Other American agencies (the National Park Service, the Bureau of Land Management, the US Fish & Wildlife Service) fall under the Department of the Interior. But even beneath that umbrella, things get complicated—for instance, selective logging does take place on National Park and BLM lands. The US Fish & Wildlife Service protects fish, wildlife, and their natural habitats—but those fish might pass through a national park or national forest via a stream, their migration blurring the boundaries of responsibility. The poaching from these conservation areas is the most shocking—trees meant to be protected through their entire life cycle and beyond brought down, a stark example of the ways in which conservation can fail.

In North America, it's estimated that $1 billion worth of wood is poached yearly. The Forest Service has pegged the value of poached wood from its land at $100 million annually; in recent years, the agency estimates, 1 in 10 trees felled on public lands in the United States were harvested illegally. Associations of private timber companies gauge the value of wood stolen from them at around $350 million annually. In British Columbia, experts put the cost of timber theft from publicly managed forests at $20 million a year. Globally, the black market for timber is estimated at $157 billion, a figure that includes the market value of the wood, unpaid taxes, and lost revenues. Along with illegal fishing and the black-market animal trade, timber poaching contributes to a $1 trillion illegal wildlife-trade industry that is monitored by international crime organizations such as Interpol.

Timber poaching is legally classified as a property crime, but it's unique in its bounty and setting. Poachers prefer the term *take* to *poach* when it comes to trees, and it is indeed that: a taking of an irreplaceable resource. In North America, trees are our deepest connection to history, our versions of cathedrals and standing ruins. When they are poached, though, they become stolen goods, and are investigated as such. But it is one thing to link a stolen car back to its owner via paperwork or plates, and another to link poached wood to the stump it once stood on. In lush forests, those stumps are usually hidden behind a curtain of trees, or covered in moss, or buried in branches—in all cases next to impossible to find.

Placing a value on poached wood is likewise complicated: the effects of timber poaching quickly become more nuanced, complex, and devastating than property crime when considered ecologically. Public lands enclose some of the oldest remaining trees in the world. Their ability to store large amounts of carbon—the redwoods alone hold more carbon per acre than any other forest in the world, and British Columbia's Carmanah Walbran Provincial Park contains twice the biomass of lush, Southern Hemisphere tropical forests that are widely considered the Earth's lungs—make old-growth trees a key species in our fight against climate change. As well, when old-growth disappears, the foundation from which it grew is destabilized, leaving landscapes more prone to flooding and landslides. Even if dead-standing (termed *snag* in the logging industry), old-growth provides an incomparable ecosystem for endangered species across the continent. When the trees disappear, so too do the animals, birds, and smaller flora and fungi that rely on them. Tree poaching, even on a small scale, has a far-reaching impact, contributing to a decline in environmental health and weakening our forests, leaving marks on the Earth that will persist for hundreds of years.

In the world of conservation-law enforcement, though, an invisible line seems to divide flora and fauna. Arguing (and fundraising) to protect animals, especially "charismatic megafauna" such as elephants and rhinos, from poaching and illegal trade tends to be easier than advocating to guard plants. But of the 38,000 species protected by the Convention on International Trade in Endangered Species of Wild Flora and Fauna (CITES)—

the global registry of plants and animals that are exploited or endangered through trade—over 32,000 are flora.

The very nature of old-growth provides an opportunity to transcend that invisible line: in Redwood National and State Parks* in California, chief ranger Stephen Troy says, the trees are "the rhino horn of the American West." The same can be said of cedar and Douglas fir ecosystems, their branches dripping with spools of moss and their trunks towering into the sky. These are trees that invoke awe, through height and age and circumference. It is very difficult to stand in a grove of *Sequoia* and not be bowled over by their beauty.

This book primarily investigates tree poaching from national and provincial parks and forests in the Pacific Northwest of the United States and Canada. These trees are only hours from my backyard in British Columbia's interior, and I have spent years trying to understand why someone might steal one. My curiosity brought me face-to-face with a form of deforestation rarely discussed, which springs from some of the most pressing social issues of the 20th and early 21st centuries.

What draws me to this story is not the amount of money that the missing wood is worth, nor even the knowledge that a single missing tree has a negative impact on climate change, though both are crucial considerations. Instead I wonder how someone who lives surrounded by the crushing beauty of a redwood forest can simultaneously love it and kill it; can see themselves as so entwined with the natural world that destroying part of it comes

---

* Since 1994, Redwood National and State Parks has comprised one national park (Redwood National Park) and three state parks (Del Norte Coast Redwoods State Park, Jedediah Smith Redwoods State Park, and Prairie Creek Redwoods State Park).

to feel like another stage in its life cycle. Timber poaching is a large, physical crash of a crime, and it is rooted in a challenge that stretches across North America: the disintegration of community in the face of economic and cultural change.

Studying timber poaching quickly opens a window into the trickle-down effects of environmental and economic policies that disregard and marginalize the working-class people who not only live among the trees but rely on them to survive. It's a difficult tale—one tinged with both anger and beauty, arising from rampant expansion and desire. The forest is a working environment, and displacing that work deprives many people of money, community, and a uniting identity. Many tree poachers express a longing for something that a tree represents: the deep-rooted underpinnings of home. The ancient Greeks called this feeling *nostos,* the root word of *nostalgia*—a searching homesickness that comes from wrenching separation.

People have "taken" wood for centuries, but wood has also been taken from us, cloistered within fences and marked boundaries on maps. Throughout history, removing land from community use often caused a wreckage, and while every poacher's story is unique, they all act out of the simmering need that followed. So why might someone steal a tree? For money, yes. But also for a sense of control, for family, for ownership, for products that you and I have in our homes, for drugs. I have begun to see the act of timber poaching as not simply a dramatic environmental crime, but something deeper—an act to reclaim one's place in a rapidly changing world, a deed of necessity. And to begin to understand the sadness and violence of poaching, we need to consider how a tree became something that could be stolen in the first place.

Chapter 2

THE POACHER
AND THE GAMEKEEPER

"Robin Hood was just taking care of his, and
his own."
                                            —Chris Guffie

"But a wild animal or a bird is nobody's property—
it's 'fair game,' and them who thinks different
thinks they own the very air."
                        —Bob and Brian Tovey,
                        The Last English Poachers

One spring day in April 1615, eleven people entered a stone
building at the edge of a forest in England's Midlands and
took their places in front of an assembled court. The group was
there to answer for their crimes: all had stolen wood from the
Forest of Corse, which they had used for things like brewing beer
and baking bread. Having been caught, they reported to this
swanimote, a court established for regulating, policing, and con-
serving the forest. In front of them sat 18 jurors; surrounding
them, 22 commoners, villagers, and farmers watched the day's

events. One by one, the accused answered for their crimes: cutting wood from pear and apple trees, lopping branches from a hazel, and in one case cutting chunks of wood from a tree known as Goblins Oak. Inklings of today's timber poaching ripple out from here.

The English word *forest* shares a root—*for*—with *forbidden* and with the Latin term *foris*, meaning *outside*. This makes sense: *forest* did not initially refer to a stand of trees or woodland, as it does today, but rather to a parcel of land that had been appropriated in the 11th century by William the Conqueror as a place where he and his compatriots could go hunting, and where others could pay for the privilege to do likewise. A sort of medieval country club, forests included more than woodland, and in some cases encompassed farmland, fields, or even entire villages or towns. When a forest was established, strict rules were placed on anyone who happened to live there: in order to preserve trees that could support a strong deer population, for instance, wood would no longer be free for the taking.

To counteract these land grabs, the 13th century brought the Charter of the Forest, a companion to the Magna Carta. Ushered in after King John, who disafforested land at the behest of wealthy barons who wanted easier access to land held tight by the monarchy, the Charter of the Forest outlined a way of life for commoners and woodlands, and allowed access to the essentials of life: food, shelter, water. *"Every free man shall agist his wood in the forest as he wishes,"* the charter proclaimed. It was a manifesto for the commons, pushing back against the spread of royal acquisition.

By today's standards, the Charter of the Forest is a radical document, standing against the privatization of common land by the powerful, be they royalty or government. The charter placed limits on use and was one of the first environmental laws in history: it included animal rights and regulated hunting with dogs. Through it, the monarchy was required to return enclosed land to its subjects. Men who had been jailed for forest crimes up until that point were released, provided they pledged never to "wrong" the forest again. For centuries, all churches in England were required to read the charter aloud to the public four times a year.

Through the charter, the forest was defined as a common source of commodities or privileges known as *mast, herbage, marl, turbary,* and *estover.* It guaranteed permission to feed pigs from the forest floor (*mast*), to let sheep graze on *herbage* throughout, and to harvest honey. It granted the right to dig clay and sand (*marl*); to mine coal and peat (*turbary*) for fuel; and to build sawmills. The forest thus outlined was a place of refuge, with trees used as sanctuary, as waypoints, and as boundary markers—there was an acknowledgment that trees were an integral part of the commoner's life, and the forest was dubbed "the poor's overcoat," under which all means of survival could be found, including dead wood or entire trees from which to build houses, furniture, doors. The Charter of the Forest also outlined the bounds of *estover*—the right to collect firewood and timber for everyday needs. It referred to coppicing, a form of logging that cuts trees down to ground level, encouraging healthy regrowth.

By the time of the swanimote meeting in April 1615, however, the charter had long since been ignored—indeed, its promises had never been fully kept. The commons had dwindled through continued enclosure of private land, primarily by wealthy landowners who removed access to communal use. Even the word *commoners* had lost its power, becoming something of a pejorative instead.

As a result of these trends, taking wood had become a folk custom by the 17th century, and timber poaching had emerged as the most common form of property crime. Forests were now a place of folk crime, where estover was routinely exceeded and trees illegally harvested and made into charcoal. "Foresters" became "gamekeepers"—de facto security guards over private property that was formerly open for common use. (Whereas the story of Robin Hood has him dodging the Sheriff of Nottingham, in reality he most likely would have been slipping the grasp of a gamekeeper.)

Keepers used methods such as snares, tripwires, and mantraps concealed in hedges to keep poachers out. Anyone caught taking wood from private land—not only in the form of trunks and branches, but also fences, posts, and bark—would be punished, and cruelly at that: seven years in prison, or hands severed, or death by hanging. Poachers were sentenced and fined by "verderers" (who held lifetime appointments) at swanimotes held every 40 days. The crimes on which verderers passed judgment ranged from cutting branches to uprooting an entire oak.

Social resentment toward staggering economic inequality began to spread—it was not uncommon for tenants to be forbidden from trapping rabbits for food, only to later watch landlords arrive and kill a hundred of the animals for sport. Forest rules had also become increasingly harsh: it was illegal to take venison for food, for instance, even if the deer had been killed by wolves. And because commoners had never accepted ceding their right to take the wood they needed, poaching became a form of resistance.

In his annual notes, one landowner complained that "neither hedges nor trees are spared by the young marauders who are thus, in some degrees, calculate in the art of thieveing." Trees are noted in records as being "carried away and injured," and one forest keeper claimed that 3,000 trees had been damaged over a seven-year span. Nearby rivers were used to transport poached venison and wood, which were stowed away on boats.

The poaching of game, fish, and timber was generally a quiet, moonlit crime, conducted through creative uses of netting, traps, and bait. (Verderers punished those who poached at night more harshly than those who worked by day.) But some poachers began to stage brash forms of protest to leave a message—killing deer on estates and leaving the carcasses behind, blood seeping into the earth; storming private estates and threatening keepers. In one case, a man wearing a woman's dress led a gang of poachers in Wiltshire. Landowners reported that their trees had been depleted, coppiced beyond recognition.

Calling themselves "the Blacks," some poachers painted their faces with charcoal to blend into the night; they swore oaths of allegiance on stag horns placed on the mantels of local pubs. In response, the English government introduced the Black Act, which instated the death penalty for more than 300 offenses—among them being "disguised in the forest." Though originally intended to be a temporary injunction against widespread crime, the Black Act remained in effect for a hundred years.

Crucially, poachers had gained local sympathy in villages and towns, where taking wood and deer was seen as the crime of folk heroes. Poaching became a means of engaging with the landscape and identifying with the land—a way to undermine the royalty and their wealthy peers, to express outrage and exact revenge. A common rhyme spread:

> *All among the gorse to settle scores*
> *These forty gathered stones*
> *To make a fight for poor men's rights*
> *And break those keepers' bones*

Chapter 3

INTO THE HEART
OF THE COUNTRY

"They call it public land. That means it's my
land, right?"
                                    —Derek Hughes

When European settlers arrived along the eastern shores of
what would become Canada and the United States, they
began cutting huge expanses of wood. The trees fell like domi-
noes, east to west, clearing ecosystems that had evolved over
millennia. They needed the wood not only for their own homes
and fires, and not only to facilitate their migration west, but also
to send abroad for the expansion of industry; America, it was
written, contained an almost universal forest, one so vast it led
"into the heart of the country."

This logging was another "taking," occurring on a foundation
of theft and enclosure—colonizers forcibly removed Indigenous
peoples by enacting violence, introducing illness, and forcing
migration from valuable land. Later, when the establishment
of national parks and forests began, Indigenous peoples were
evicted from land that would become iconic parks providing
staggering views: Yosemite, Yellowstone, Glacier, Badlands.

The American project was one of expansion, a taking of new soil. Soon, five billion cords of wood had been consumed for fuel alone, cut from 200,000 square miles of woodland over 50 years—an area equal to Illinois, Michigan, Ohio, and Wisconsin combined. Some loggers saw their work as bringing the heavens closer, the crash of canopy leading to "the light of civilization being drawn upon us." Many others saw big trees as stubborn, standing in the way of expansion, obstacles to be conquered. To bring them down, loggers sometimes stuffed black gunpowder and a fuse into holes carved into trunks; when it exploded, the log would split perfectly in half. Archival photos from the time show men perched on massive stumps, with captions such as "Cut 'er Down, Boys—There's Plenty More Over the Next Hill!"

Meanwhile, the American conservation movement was born in cities as urban areas grew. It was not uncommon for doctors to prescribe time in nature to cure headaches and nervous break-downs, sending patients to rural areas to escape the noise and smell of congested city streets. As city dwellers started visiting regions such as the Adirondack Mountains of New York, they also became invested in preserving them. Because they were arriving from crowded streets, many saw the benefits of con-servation as preserving places that humans hadn't yet sullied—in reality, untouched nature never existed. Forests were logged by the working poor, who then built homes in rural areas outside the country's burgeoning cities. Now they were being forced from that land—told that their homes were subpar, that their work had caused harm, that the environment mattered more.

Conservation was lobbied for—and funded—by wealthy do-nors whose use for nature was in recreation. Organizations such as the New York Sportsmen's Club were founded and lobbied for stronger conservation measures to ensure access to game and fish that paying members wanted. They lobbied to ban the sale of game meat, and their work led to dedicated hunting and fishing seasons. Fishing with nets—a common practice used by farmers and rural residents to catch large amounts of fish to feed their families—was banned.

As it had in England only a few centuries earlier, hunting became poaching. Foraging and grazing became trespassing. Logging became timber theft. Shortened hunting seasons were dictated by sport hunters who didn't care to consider the cycle of harvest seasons, making farmers choose between tilling the land and hunting for meat at certain key times each year. "When you say to a ranchman, 'You can't eat game, except in season,'" one Wyoming man wrote to his local newspaper, "you make him a poacher, because he is neither going hungry himself nor have his family do so.... More than one family would almost starve but for the game."

In 1892, when Adirondack officials began mapping out the park's official boundaries, they noted that many locals were unaware of the distinction between parkland and their own homesteads. Lines dotted on maps were not marked in the forests themselves, leaving unintentional trespassers in the lurch and in some cases transforming settlers into squatters. Even if a settler had lived in their home for decades, they could be forced to leave if a park was established on the surrounding land; in some cases, park commissioners argued that the squatter was "...not a desirable neighbor. His abode is not only an eyesore, it is too often surrounded with a litter of old tin cans, fish scales, offal, hair and hides...." As rural residents were evicted, often by tearing down and burning structures, an environment of resentment and revenge was born in places such as Pennsylvania, upstate New York, Virginia, and Vermont. A National Forest Commission, which included the preservationist John Muir, eventually recommended that the army be asked to patrol reserves. When President Grover Cleveland set aside 21.3 million acres for new forest reserves and parks in 1897, business interests and politicians in the western states were incensed. They saw it as "useless protection of dead timber" and dismissed these protectors as "zealots, Harvard professors, sentimentalists, and impractical dreamers."

Timber poaching became a frontier tradition for some settlers. In the Adirondacks, locals began to rip down NO TRESPASSING signs before tromping into the forest, often to harvest wood. The transgression was incredibly difficult to prosecute: wardens relied on tip-offs from locals to catch prolific poachers. "It is almost impossible to obtain evidence against any individual in a locality unless there is some man in that section who has some ill will against him," wrote an inspector for the Forest Commission. "If they tell what they know in regard to these trespassers to the State officials they subject themselves to the annoyance and to the ill will of their neighbors and it makes life unpleasant for them at home." Some rural residents simply didn't see their actions as poaching; when caught, many were so angry that they retaliated by setting the woods on fire.

Revolt continued into the 20th century. In September 1903, a landowner was shot to death after suing a local man for poaching wood on his property. The landowner had purchased the rights to a local road and then blocked access to it, and he had bought a stream previously used to float logs to a local sawmill. Private lands became the subject of rural ire: local residents burned down estates, cut holes in fences, fired guns at guards. William Rockefeller began to travel with armed bodyguards, and bullets were fired into his lodge at Bay Pond. His guards began to quit.

Indigenous peoples, too, continued to take plants and animals for subsistence. Theirs was a deeply rooted rebellion against settler influence—they were true locals, with an intimate knowledge of the land, "taking" back after America's original, founding theft. Poaching was a subversive method of reasserting traditional rights and practices. In northern Canada, Chipewyan hunters protested the establishment of buffalo preserves, and continued to hunt and trap in Wood Buffalo National Park after its 1922 creation—and were punished harshly for doing so.

Wardens during this era worked in dangerous conditions, and some were killed when confronting hunters and poachers. They reported that local residents were brazen in going into the forest for firewood: "The people around the borders of the wilderness had been educated from time immemorial, that is, from the first settlement of the country, that what belonged to the State was public property, and that they had a right to go in there and cut as they wanted to; their fathers and grandfathers had been doing that, and that they had a birthright there that no one could question." Today's most iconic parks were embroiled in these revolts. Yellowstone had a famous poacher, named Edgar Howell, who wrote letters to the local newspaper arguing that he hunted in the park because it required skill and bravery. He proclaimed it a "rush" to outwit rangers at their own game. In response, conservationists dubbed him worthless and greedy.

Writer Donald Culross Peattie credits the "first assault" against California's redwoods to the gold rush in 1850. Though Spanish settlers had set foot in redwood forests in the 18th century, a hundred years later the final hill that those photographed, ambitious loggers from the East would pass over led them straight into Humboldt County and the shores of the Pacific Ocean.

Humboldt County, named for German geographer and scientist Alexander von Humboldt (who never visited the region), is 270 miles north of San Francisco. On arrival, loggers stepped

onto the territory of about half a dozen Indigenous groups—
the Wiyot, the Yurok, the Hupa, and the Eel River Athapaskan
peoples, to name only a few. The region's forests were filled with
old-growth redwood and spruce that had evolved over millions
of years to suit the coastal mountain range. The landscape had
sustained Indigenous peoples for millennia, who commonly built
long, low, wood-plank homes, secured with grapevine bindings,
along riverbanks. The planks that formed these homes were thin,
pliable pieces of wood that were split from naturally falling or
standing redwood trees. Canoes were carved from large, fallen
trunks. Sustainable use was embodied in a traditional Yurok story
explaining how redwoods, or *keehl,* had sprung from Creator to
be used as boats and houses—the redwood was a living helper.

Redwoods fall under the genus *Sequoia* and are part of the
cypress family—the Cupressaceae. The world's tallest trees, redwoods are true relics: they have grown on Earth for 100 million
years, at one point taking root as far north as the Arctic. All of
the Pacific Northwest is rugged, soaked in rain between October
and May, and stacked with enormous trees: some have trunks six
to eight feet in diameter, and they routinely grow more than 250
feet tall. Stretching upward from the shore, redwoods, coastal
Douglas fir, and the shrub toyon (thick with red berries in fall)
carpet the hills and mountainsides. A redwood's canopy widens
only to the edges of its neighbor's crown, so as not to crowd
each other out. With your head tilted back and eyes turned
upward, you see deep rivulets of sky that stitch between crowns
like rivers. From the ground, the sky is not the canvas, but the
thread. This is what loggers walked into, and what still exists
today: an ethereal and ancient forest.

In the 1850s, these redwoods seemed endless. Dubbed "red
gold," the trees were deemed as important as minerals and
turned into houses, stores, sidewalks, sluice boxes for gold

panning, casks, ships, and even broom handles at local mills. At the time it was estimated that there were two million acres of redwood forest in northern California, encompassing wide, rushing, aquamarine rivers, four large watersheds, and an ecosystem rife with salmon, sea otters, and birds. Redwoods undulated to the horizon, a blanket of lush green hills. "Trees! Such monsters, all crammed together as thick as corn stalks," wrote pioneer Amantha Still in her journal in 1861.

Small towns were being cut from swaths of this timber, many of them at the confluence of forest and ocean. One town, Orick (derived from the Yurok word for *mouth of the river,* though a competing founding story has settlers hearing green frogs singing *O-rek, o-rek*), thrived in a lush, rain-soaked valley that provided good land on which to raise dairy cattle. Some community members in Orick today can track their ancestors back to this dairy boom. "When my mother was born here in Orick, there was no timber operation," says Ron Barlow, a rancher in town.

Some of North America's oldest timber companies were founded not far from Orick. In the 1880s, the Eel River Valley Lumber Company estimated that it could produce 7,500 shingles per day for about 20 years before running out of raw material. In the late 19th century, it was common for whole houses, banks, or churches to be built from a single redwood. (The tree is that easy to work with: pin-straight, smooth, and light on resin.) And early in the 20th century, almost every city in the United States used redwood pipes to transport its water. Brewers in Milwaukee used redwood vats; mining communities in Utah used redwood flumes; and to this day, some electric water heaters still have redwood insulation. Orick's first commercial sawmill opened in 1908. It milled redwood, but also spruce trees that had earned a reputation for being beautiful and of high quality—often eight feet in diameter, and stick-straight, so perfect for milling.

North of California the redwoods thin, giving way to a temperate rain forest rich in Douglas fir, balsam, Western hemlock, red and yellow cedar, and spruce. In the shadows of these trees, thousands of communities were built along riverbanks in the early days of colonization, most of them dubbed "stag camps" because they were populated primarily by single men. In Washington and British Columbia, trees grew in dense forests all the way from mountain slope to shoreline, and logs were eventually sold to the eastern United States and Europe in great rafts, which were floated downriver and shipped from San Diego. At the turn of the 20th century, timber company Weyerhaeuser bought 900,000 acres of thick northwest forest, one of the largest land transfers in United States history, for $5.5 million. Some of the most productive forest in the world stood along the southwestern shore of Vancouver Island, where the boundaries of Carmanah Walbran Provincial Park are now.

The Pacific Northwest offered something that logging across the rest of the United States could not: stability. The story of westward expansion would become one of heroic men and their enterprising bosses, felling trees with "tiny tomahawks," then massive saws, then chain saws, then powerful and efficient machinery. Timber companies began spreading the lore of Paul Bunyan, a lumberjack folk hero dressed in plaid, whose prodigious physical feats in the woods quickly became the stuff of legend. Imagery of Bunyan and his exploits began to appear everywhere loggers went. Here is the logger's identity, he advertised, of which you are a part: masculine, independent, skilled, solitary.

Bunyan confirmed a feeling many held deep in their bones: in their migration, many Humboldt settlers had survived extreme weather, the deaths of children or entire extended families, drought, and shipwrecks. Once they made it to Humboldt and set up homes in towns like Orick, a narrative centering on toughness settled in—they had *made it* here. "In no part of the world is there to be found a more efficient or hardier class of men than are occupied in the redwoods," declared the *Humboldt Beacon* in 1913. An identity was shaped: productive, concerned about today and not tomorrow, each logger his own man in an isolated camp in the big woods. Some lone woodsmen even lived in "goosepens," or hollowed-out redwood trunks large enough to house a grown man.

The first notions of redwood conservation track back to this time. In 1915, the president of the National Geographic Society, Gilbert Grosvenor, traveled west to document *Sequoia* and capture the forest in photographs. Two years later, three conservationists—John C. Merriam, Madison Grant, and Henry Fairfield Osborn—embarked on a road trip that led to the formation of the Save the Redwoods League.

The trio of Merriam, Grant, and Osborn drove the future Redwood Highway to see the trees that produced stumps wide enough for entire communities to pose on, for whole ballroom floors to be crafted from, for literal political "stump speeches" to be dictated from. By then the English-American businessman William Waldorf Astor had already purchased and shipped to England a circular slab of trunk, a cross section from a redwood 35 feet in diameter and estimated to be 3,500 years old; from it he would fashion a large dining table to settle a bet. The trio was appalled at the open, zealous logging that greeted them in California's northern reaches. They were also eugenicists who saw parallels between environmental destruction and the decline of Nordic supremacy. They considered protecting the redwoods as part of a mission to enshrine a White, masculine dominance over the wilderness. After the three men formed the Save the Redwoods League, in 1918, their observations, alongside

Grosvenor's photos, encouraged private, wealthy individuals to buy groves and preserve them. Puzzle pieces of land were slowly preserved as state parks—small enclaves surrounded by terrain still being logged.

They were soon joined in their mission by Newton B. Drury, a close adviser to the multimillionaire financier and oil baron John D. Rockefeller Jr. Drury defined himself as "the person who had to bear all the slings and arrows of both good and outrageous fortune" in eventually establishing what would become Redwood National and State Parks. His name is emblazoned on signs almost everywhere in the region, and driving through the park now, you will follow the Newton B. Drury Scenic Parkway. "The primary purpose of the national park," Drury said an interview, "is one of resisting…any attempt to turn to utilitarian purposes the resources represented by the forest."

Theirs was a resistance facilitated by the wealthy: It germinated in the halls of governance, in parlor rooms, in private meetings. It relied on privilege and access to power. Drury hosted grand picnics beneath the redwoods for deep-pocketed donors and powerful lawmakers. Their activism would enshrine in practice a form of conservation that was only marginally based in the region itself—one that required lobbying for hundreds of millions of dollars to buy swaths of land from private logging firms. The process of garnering support for the park was conducted seemingly blind to those who lived and worked there; many large donors resided in the East and saw conservation not as a practice of "wise use" (pioneered by the Forest Service's first head, Gifford Pinchot) but as preserving a landscape untouched by human hands. Theirs was a conviction held by the likes of President Theodore Roosevelt, who upon visiting the Grand Canyon in May 1903 had declared, "Leave it as it is." (Implicitly: wild, empty, devoid of humans.)

To carry out this mission, though, the Save the Redwoods League had to hire a logger. The League contracted the trusted northern California woodsman-polymath Enoch Percival French to "cruise" the woods and provide an accurate estimate of the remaining old-growth redwoods. His was some of the first work done to measure the amount of wood available in the forest—a number that would prove both the ecological and economic power of preserving the woods. Because of French's cruises, we know that whereas the average forest produces about 30,000 to 40,000 board feet per acre, the redwoods produce 60,000 to 65,000. In the region of Bull Creek, where the stand is 100 percent redwood, that figure spikes to 200,000 board feet per acre.

French also bridged the worlds of conservation lobbying and boots-on-the-ground experience. He understood that logging had become infused with a kind of moral value that rested on the ideals of hard work, independence, self-sufficiency, and skill with an ax. By 17, he had begun to work for the Pacific Lumber Company alongside his father. Strapped for cash at the time, French knew there would be plenty of space between him and his bosses once he got into the woods: "I could go out and cut any tree I could find," he remembered. "So I went out now and then and cut eight or ten trees for split lumber. They never missed them." French would turn the logs into shakes and sell them for use along the railroad, netting him about $4 per thousand board foot.

In 1931, French became the first ranger of Northern California Redwood State Parks. Every morning he drove through the parkland until the road grew too rough, then hopped out to walk the park boundaries. When roads washed out during the rainy season, he rode logs down inflated rivers, steering with a paddle made from downed wood. When the Ladies Garden

Club from Del Norte County, to the north of Humboldt, showed up for a visit before donating money to the park, French carried them piggyback over a river, one at a time. ("They didn't seem to mind," he recalled in a 1963 interview with oral historian Amelia Fry.) The garden club was composed of financial donors, who requested that a permanent pond be formed and preserved in their name. French had to explain that this was impossible: redwoods can't survive in standing water. Privately, their request frustrated him. There seemed to be a misunderstanding, he said, between what was natural to the forest and the aesthetic beauty that some wanted to exist in nature.

French had turned from lumberman taking wood on the sly to park warden responsible for protecting that resource. By the time he became a ranger, a redwood log fetched up to $100 per thousand board feet. In his 20-year span as a ranger, French estimated that timber poachers had taken out the equivalent of two to three million board feet of timber, as well as ferns and lilies that grew in the undergrowth. "I knew all the boys," he said. "I don't like to mention their names. I was raised there myself...."

And French knew something else: he too had been a poacher, earning extra cash selling railroad shakes. Still, he eventually decried their work: "They were out killing deer in the parks. To me that didn't hurt anything, a deer is all right. But if a man's out there with a truck and he takes some trees off, it takes 500 to 1,000 years to grow them again....

"Anyhow, that's what I was there for."

Later still, Enoch Percival French would brand the practice of timber poaching "pitiful."

Chapter 4
A LUNAR LANDSCAPE

"A lot of times they're going to describe themselves as an out-of-work logger...when maybe
their parents were out-of-work loggers."
　　　　　—Phil Huff, Forest Service special agent

Enoch French's redwood activism was rooted in his ancestry. His father, a logger, believed in preserving the redwoods as much as he believed in his right to harvest them. French trusted in the regenerative power of the forest—that sometimes great destruction can lead to new growth that brings immense beauty. Floods, landslides, the trampling of undergrowth: "That's just nature's way of improving things," he told historian Fry, "if you really want to get at the truth of it."

French's time as a redwoods ranger overlapped with a working-class environmentalist movement in which worker and environment were allied against overextraction, not each other. His sentiments are echoed in interviews from the early 20th century. Logger Charles E. Hunt, for example, said that loggers chose their profession so they could live in the woods: "Maybe no logger could put it exactly into words, but he stays hard at work in the forest because he loves trees."

In the shadow of the Great Depression, President Franklin D. Roosevelt deemed clear-cut logging "a matter of national concern," and many logging unions began to advocate for conservation. The president of the International Woodworkers of America (IWA)—a Canadian shingle weaver and Communist named Harold Pritchett—went on the radio in Seattle to explain what forest conservation would mean for workers: more consistent employment over a longer period of time; an opportunity to reforest land that had already been logged; and a commitment to the region's future that a corporate "cut-out and get-out" policy had not allowed. In particular, Pritchett insisted, the IWA wanted everyone to understand "the work humans did in the forest and the work forests did for humans."

In the wake of the Second World War, however, a housing boom and increased demand for pulp and paper would challenge forest conservation and lead to heavy logging. It was part of a "conspiracy of optimism," during which time wood, the government said, was the country's most important asset: it would help rebuild better lives at home. This was a time of industrial revolution in construction, an expansion of scale, and a call to make the country great by building close to five million new homes. The boom produced record cuts, high employment, and the emergence of clear-cutting as the dominant mode of harvesting. It also left a legacy of environmental destruction that would spark more calls for conservation.

During the boom, the tiny town of Orick, California, welcomed loggers and their families to the surrounding hills, watching its population eventually swell to 2,000 and its number of sawmills to four. School class sizes expanded commensurately, and more teachers were hired. Some of the logging firms paid so much tax that the community could operate solely off their success. It was a community in flux—the highway was lined by neat rows of motels, which some remember being frequented by a "continuous stream" of logging rigs. "All Orick needs is time," one resident told a reporter at the time. The riverbed was home to a makeshift encampment, where entire families lived in tents. "Some of our people are living in hollow trees and under old boards now, but every town goes through that in its boom days. Come back in a couple years."

The influx brought a man named John Guffie to town. When Guffie had started logging—taught alongside his brothers by his father—he was told that if he learned how to be a good logger he'd never be out of a job. He had grown up in western North Carolina with nine siblings, and logging was so much a part of his life that he figures he learned it by osmosis. "That's where you get your ideals from," he explains. "It's a life experience."

Eventually, however, logging jobs did start to die out in North Carolina, and in 1955 he moved to Humboldt County, following a brother who had found logging work there. Guffie got married and had three sons and a daughter. His wife, Kitty, was known for being strong—though scarcely 110 pounds, she was regularly seen slinging a sledgehammer. Guffie's work as a logging supervisor migrated from one big Humboldt timber company to the next: Hammond Lumber Company, then Georgia-Pacific, and finally Louisiana-Pacific, a Georgia-Pacific spin-off. He took photos of his kids in diapers, sitting on logging equipment. He signed his sons up for Pop Warner football. He became a preacher and officiated weddings.

But not long after John Guffie arrived, the town's geography would be altered forever.

The May Creek watershed helps stitch together the large, arterial waterways that create the redwoods' Prairie Creek and Redwood Creek basins. These waterways act as the lifeblood of Humboldt County's ecosystem—a redwood relies on water as much as it does on the earth in which it is rooted. Redwood trunks reach straight up into the sky, and the fog that curls along the craggy coasts keeps their foliage hydrated. Because coastal redwoods are so tall, water absorbed by the roots often can't reach the crown high above. In dry months the redwoods rely on fog as they might a strong downpour, their leaves sucking in moisture and nutrients such as nitrogen. This leaves the tree's roots free to store groundwater, which prevents riparian zones from drying out. Even in times of drought, logs decaying on the forest floor often feel damp, and they furnish water to organisms throughout the forest. But there is strong evidence that forests don't thrive when they are consistently trafficked or disturbed beyond the rhythms of nature. And the redwoods were heavily trafficked.

During clear-cut logging, topsoil is lost and streams are bull-dozed for roads. By the time Guffie arrived in Humboldt, this process was taking place in the redwoods on an epic scale. In Humboldt's forests, the root system could no longer contain the immense annual rainfall, and the waterways began to flood. Mangled roots, lack of second growth, and flattened shrubs made the earth unstable, and the construction of roads deep in the woods to transport logged wood had hastened erosion and habitat destruction. In December 1955, heavy rains drenched the north coast of California—upward of 24 inches in three days—and high winds snapped branches and tree limbs, which were carried in rough currents down mountainsides and into the town of Orick.

As the rain continued to fall, the weight of the earth soon outstripped the ability of the land to contain itself, triggering a landslide that toppled 1,000-year-old redwoods and covered the region in silt and mud. One resident remembered seeing entire houses "lashed to logging trucks" to prevent them from floating away. A ranger in the area described the forest in the aftermath as "a lunar landscape, with the raw edges of the mountains exposed."

Logging didn't slow after the catastrophe, though. In 1964, rains once again pummeled the high ground, and another powerful flood swept through the town. This triggered a sense of urgency among environmental groups, including the Sierra Club, which, along with the Save the Redwoods League, wanted to institute conservation measures that would stall clear-cut logging in the region. The floods supplied the strong visual and emotional representation of the negative effects of industrial logging the groups needed to get a national park established.

Soon after that second flood, and about a decade after he moved to Humboldt, Guffie says, he went to a meeting at Louisiana-Pacific and heard a proposal for turning some of the company's holdings into a national park. His back was up. "I thought, just like anything else, it's just another politician planning for themselves and not planning for the people . . . when they're taking your job away and telling you they're doing you a favor."

Lobbying to establish a park had begun in earnest. Hundreds of statements were recorded during an April 1968 tour of northern California by the US House of Representatives Subcommittee on National Parks and Recreation. On April 16, Orick Motel operator Jean Hagood sat before the subcommittee chairman and posited that a national-park economy fueled by tourism would be more sustainable for her town than boom-and-bust timber. But she was one of only a few locals to hold a candle for conservation. When members of the Sierra Club met with local supporters in Humboldt, they prudently parked their cars blocks away, careful to maintain the privacy of residents loath to alert neighbors they sympathized with the cause.

The proposals to transfer timber cutblocks—areas that had previously been authorized for harvest, some privately held, some managed by the Forest Service—to the National Park Service were not popular. At the same hearing where Hagood voiced her support for a park, her neighbor Mary Lou Comstick argued that closing the two mills near Orick would damage the town's farms and dairies. Many local trade unions protested vehemently against the park; they sent members, including rancher Ron Barlow, to picket public-park consultations with signs bearing the slogan DON'T PARK OUR JOBS or wearing T-shirts emblazoned TIMBER FAMILIES: AN ENDANGERED SPECIES. A job as a janitor in a park building, they said, could never provide the income and opportunity of a logging job. The subcommittee heard warnings from a forester from Washington state, who characterized Olympic National Park (a 900,000-acre park designated in 1938) as a harbinger of what was to come: "The towns and communities surrounding the park have shown a much lower economic and population growth than the state as a whole," he cautioned.

Confronting all this opposition, the US Department of the Interior proposed financial bailouts for those in affected communities such as Orick. The Save the Redwoods League advised that logging communities should be compensated for their lost taxes and given government relief on a diminishing basis, until tourist travel could offset their disappearing industrial revenue. But the league's own executive director, Newton Drury, later admitted that "it's a grave question whether [logging communities] would ever recoup, entirely. There's a limit—and there should be a limit—to tourist travel in the area." In turn, the Orick Chamber of Commerce requested that no food concessions or services be built inside the park; that way, visitors would have to spend money in the town.

Over a two-year span, the park bill made its way through Congress, eventually preserving 58,000 acres of redwood land, including some 18,000 acres along Redwood Creek and its drainages. Redwood National Park was officially designated on October 2, 1968, its boundaries cutting right along the edge of Orick. As requested, no kiosks or campgrounds were built within the park boundaries, meaning visitors would have to fill gas tanks, pitch tents, and stop to eat in town.

The historical legacy of the founding of Redwood National Park often focuses on the precedent set by a federal government annexing private land and taking it into public ownership, rather than on the fallout for communities in the region. Though corporations were compensated for their lost profits, direct government relief for workers never materialized. Mills shut down, and logging companies fled the area after the park shuffled their holdings. The remaining economy was primarily service-based: gas stations fueling tourists' cars, Jean Hagood's motel offering them a place to sleep. Lady Bird Johnson dedicated the park the following summer, and a grove of old-growth redwoods was named after her.

Logger John Guffie launched his own timber company, carving out a space for work in the remaining forest.

While some pinpoint 1968 as the year Orick's economic troubles began, it was only the start of a slow change that unfurled over the following decades, sowing the seeds of chronic unemployment, housing decline, and anti-establishment sentiment that smoldered before erupting across the Pacific Northwest in the Timber Wars of the 1980s and 1990s.

Along with the national park, the late 1960s and early 1970s brought an influx of newcomers to northern California. In the burned-out shadow of Haight-Ashbury's hippie dreams, many people headed north as part of a countercultural movement. They found dying logging hamlets, lots of room, and community in cities such as Arcata and Garberville. Many of these new residents were credited with "parting the Redwood Curtain"— that is, easing the region's cultural isolation by bringing the outside in. Southern Humboldt County, soon dubbed "SoHum" (and then, inevitably, "Shum"), became a promising location to stake your place in the new agrarian movement. Writer David Harris observed that two types of people lived in Humboldt: "Those who looked like they had just got out of the Marines, and those who looked like they had got out of a Grateful Dead concert."

In her book on Humboldt, former Berkeley political activist Jentri Anders (who lived in Shum and submitted prolific letters to the editors of local newspapers from her vantage point in the counterculture) wrote that Humboldt at that time was undergoing a "huge crack." With northern California going through an economic exodus as mills closed, back-to-the-landers in Humboldt and Mendocino Counties were able to settle "whole watersheds previously undeveloped." When they arrived, they witnessed how industry affected the natural world they had hoped to escape to. They found their water supplies polluted by pesticides that had been used on logging land, and they lamented the rampant capitalism surrounding logging, which was antithetical to their priorities. To Anders, the melding of hippies and timber workers was "muddy"—a new community that defined itself under the counterculture, attempting to integrate itself (and its ideas) into a previously established community that thrived and suffered at the whims of natural-resource extraction.

At the same time, many environmental activists remained unhappy with the amount of forest that had been preserved by the 1968 creation of Redwood National Park. In 1976 the Department of the Interior proposed further expanding the park to protect a 48,000-acre parcel of land at the top of Redwood Creek that was due to be logged soon. The proposed expansion would increase the park's size to 106,000 acres of protected forest, drawing together a patchwork of smaller parks that had, for the past decade, floated as islands. It would also allow the rehabilitation of previously logged lands.

The logging communities of Humboldt County found themselves facing the prospect of yet more shuttered mills. More than 1,300 people were expected to lose their jobs if the park expansion went through, including 611 who worked in logging and at mills. And just as it had failed to do in 1968, tourism could not be counted on to fill that gap: while more than 400,000 tourists were estimated to visit the Redwoods parks each year in the 1970s, many of them drove straight through towns such as Orick without stopping; instead they parked their cars at Redwood Highway pull-off points and trailheads for short hikes.

Timber companies, truckers' unions, and loggers pushed back against the expansion. (JOBS DON'T GROW ON TREES was one popular slogan on the signs that began to proliferate around the region.) At public events, speakers implored the park and government representatives to "work for the general welfare— don't put us on welfare." William Walsh, an executive with the investment firm Arcata National Corporation, testified to a US Senate committee that "the cold and the rain and the fog which are so essential to redwoods make the area unattractive to vacationists."

Once again, the region's remaining logging firms were offered financial reimbursement from the federal government to alleviate the effects of park expansion. This time around, additional funds were allocated to retraining, job retention, and community economic development. The government committed to investing $33 million in watershed-restoration projects, with a promise to hire former loggers and mill workers to do the work. A further $25 million was designated for the Redwood Employee Protection Program, which provided income and benefits to those who lost their jobs in the industry. The Forest Service was asked to consider increasing logging in the nearby Six Rivers National Forest. Notably, the Interior Department was also directed to fill 60 new park positions with individuals who had lost jobs as a result of the park expansion.

None of these offers did much to assuage the fears of logging families. Interior Secretary Cecil Andrus predicted the social fallout from the park's expansion: "The problem we will have will be with the individual. Let's say someone [is] 50, 55 years old, and has been a cat skinner in the woods for most of his adult life. He has lived in one of those small towns in that area his whole life, [and] he is going to be very difficult to train and move."

By 1977, a rising tide of Humboldt loggers felt they weren't being heard by the Department of the Interior. In response they organized a convoy of trucks to drive from Eureka to Washington, DC—a trip that would spread their message independently from a media they felt favored the conservation effort. Residents in Orick held spaghetti dinners, barbecues, and raffles to raise money for the trip. They felled a dead-standing redwood and loaded the logs on the backs of trucks: "It was to show people this is dying; this thing in a few years is going to be on the ground—wasted, gone," explains Steve Frick, a retired logger, from his Orick home.

The "Talk to America" convoy would be led by a red semi-trailer hauling a 19-ton, 19-foot-long chunk of redwood that had been carved into the shape of a peanut. The end goal would be to present the sculpture to President Jimmy Carter—the peanut farmer. IT MAY BE PEANUTS TO YOU, BUT IT'S JOBS TO US! read a sign on the rig. HOW MUCH IS ENOUGH?

The procession set out from Eureka in May 1977, joined en route by truck drivers from Washington, Oregon, and Alaska. The trip took nine days; along the way they stopped in Reno, Salt Lake City, Detroit, and other cities, where the loggers convened downtown to hand out redwood saplings and make their case against the park. But the highway brought plenty of pushback. The convoy's trip was punctuated by people flashing their middle fingers and cussing at the drivers. "They disagreed with the whole thing," says Frick. "They thought everything should be locked up tight."

Once in DC, loggers in their work wear and hard hats assembled on the steps of the Capitol Building and held a protest. Trucks pulled up outside the Capitol and sent word that they had a gift for the president. They parked the truck with the redwood chunk on it nearby, and someone turned the sprinklers on them. Carter dispatched two aides, who listened to the loggers' speeches but declined the gift of the giant peanut, calling it inappropriate and a waste of America's precious wood. "There's been an impractical use made of this," Carter's special assistant Scott Burnett told the group. "We'd like to see something practical done with it."

That's how Frick found himself at the wheel of a logging truck on the drive home through Colorado, "mad as hell and tired," as a Volkswagen van pulled alongside him on a highway. The van's passengers gave him the finger and started yelling out the windows. Frick normally would have let them pass, but the truck driver ahead hailed him on the CB radio: "I want them—get them stopped." The pair pinned the van between them, blocking its way and forcing it to speed.

Frick's truck began to slide into a ditch. His wife, in the passenger seat, started to scream. So the pair of loggers eased up, letting the van go. At the bottom of a hill, however, the convoy caught back up with the van, now parked beside a telephone booth. The driver ahead of Frick parked his truck, got out, and walked to its back frame, then used it to swing himself up and drive both feet into the van's windshield, splintering it. The convoy found themselves with a police escort all the way out of Colorado.

In the end, the park expansion went ahead despite objections. Groups such as the Sierra Club and the Save the Redwoods League had capitalized on public guilt, drumming up enthusiasm among their urban supporters for further protecting the redwoods. Forest sociologist Robert Lee says city dwellers are more likely to feel guilt toward nature, which he attributes to disconnection from nature rather than empathy toward it: "They are very likely to regard trees as a symbol of immortality or continuity," wrote Lee in one study. Rural residents, by contrast, "can live with the ambivalence of loving nature and cutting trees. It's an acceptance that that's life."

"People almost regard [redwoods] with religious feelings," Sierra Club director Edgar Wayburn later told historian Amelia Fry. "I think this factor more than anything else allowed us to put [the park's expansion] over."

The redwood peanut still sits outside the Shoreline Deli and Market in Orick, slowly rotting in the rain and crumbling into the ground. A reminder of a fight.

Chapter 5

REGION AT WAR

"Hell yeah, it hurts. They took everything away
and then we lost it all."   —Chris Guffie

In October 1982, Derek Hughes was born in Sparks, Nevada, to Lynne and Dennis Hughes. His parents divorced when Derek was a toddler, and he and his older sister moved with Lynne to Sacramento, where she had extended family. While there, Lynne met and married Larry Netz, who would go on to help raise her young kids. The couple wanted to live somewhere with an affordable cost of living, and in 1993—the year Derek entered sixth grade—they moved the family to northern California, settling down in Arcata.

More than a decade had passed since the expansion of Redwood National Park, and a study from the US General Accounting Office in the early 1990s reported that the economic and employment programs introduced in Humboldt County in response had not worked. The report said that many people took advantage of benefits for which they did not qualify, and that benefits may have discouraged workers from finding new jobs. Retraining programs had been delayed. By 1988, $104 million had been spent on about 3,500 people, of whom fewer than 13 percent had received retraining. Whatever economic recovery had occurred in the region was attributable to an influx of retirees. "Never have so many given so much for so few," one critic noted of the funding.

The report confirmed what the past decade had borne out in the Pacific Northwest, which had entered an age of economic upheaval. In the two decades following the park's expansion, a battle dubbed the "Timber Wars" would spread up and down the Pacific Northwest. In Canada it became known as the War in the Woods, climaxing in a showdown in Clayoquot Sound on Vancouver Island, and with protests on the island chain of Haida Gwaii. Derek Hughes had arrived in Humboldt County just in time to see anger roil over the region.

During a recession in the early 1980s, a slump in demand for building materials led to economic turmoil and layoffs in logging communities. Oregon's unemployment rate reached 20 percent in 1982; logging companies across the region broke union agreements and cut hourly rates. Families were thrust into economic uncertainty, as well as social instability: in 1983, a study of unemployment and its consequences reminded readers that "statistics are people."

The economy had only just begun to recover from that recession when, in 1990, the northern spotted owl was designated a threatened species under the Endangered Species Act. The logging communities in Cascadia,* which shared a bioregion with these tiny birds that are used as an indicator of ecosystem

---

* The Columbia River watershed and the area around the Cascade Range. The Cascadia bioregion stretches between coastal Alaska in the north and northern California.

health, were turned on their head. The northern spotted owl requires large tracts of old-growth forest to survive, and the depletion of old-growth due to clear-cut logging in the 1960s and 1970s had gravely imperiled the species. Though it was common for logging companies to commission tree-planting projects to encourage new cycles of growth, the spotted owl serves as a prime example of why such tree-planting efforts don't make a forest. Most often, second-growth land is planted over with a single tree species such as Douglas fir, which grows fast and can be harvested relatively quickly. But the spotted owl (as well as other species, notably the marbled murrelet) live in old-growth only: they prefer to nest in burrows made in wide-circumference trunks, and very tall trees make the perfect environment for hunting prey from above.

Logging that would degrade the habitat of the spotted owl was forbidden from that point forward. Any logger who saw an owl flit between branches was duty-bound to report it, whereupon all work would stop. The bird became the mascot for activism in Washington, Oregon, and northern California—it represented how the forest *should* be, if only it was left alone. And its classification as an endangered species was a boon for groups such as the Sierra Club, which began using lawsuits (alongside protests and lobbying campaigns) as a tactic to stop logging. Such suits—against the Forest Service and private corporations—suspended logging until their outcome had been decided. This, argues labor historian Erik Loomis, "undermined potential allegiances" between activists and loggers, the latter being the ones most affected by lack of work.

At the same time, a similar battle was taking place on Vancouver Island, in the forests that would eventually become some of British Columbia's most iconic parks. In the mid-1970s, the province's Forest Practices Act had handed over control of much of the island's remaining forest to a handful of corporations. As clear-cut logging intensified into the 1980s, vast environmental damage resulted. During fieldwork, for example, forestry workers noted that clear-cuts were "like huge ovens"—very hot, and not at all conducive to new growth. "Some of us recognized a decade ago that too much is being cut," one logger said at the time. "But no one cared. We had no clout, and our warnings fell on deaf ears."

In April 1993, the provincial government released a plan for logging in the Clayoquot Sound region. Two-thirds of the region's old-growth rain forest would be open for public logging contracts. In response, protests spurred by environmentalists escalated throughout the summer, with an estimated 11,000 people participating in a five-month protest that stretched into the fall of 1993. It would become the largest act of civil disobedience in Canadian history.

The War in the Woods occurred at the confluence of shocking environmental destruction and devastating job loss: 23 percent of jobs in logging were lost from 1980 to 1995. At the same time, production was on the rise, and clear-cuts pockmarked the forest. Workers and environmentalists were pitted against one another. Task forces across industries and interests were formed but often quickly dissolved—in some cases, environmentalists walked out of negotiations because logging continued while discussions slowed. There were counterprotests by loggers, who told the media they wanted their children to have an opportunity to work in the woods. Some companies engineered massive walkouts; 15,000 timber workers took part in these all told, the largest mass protest in British Columbia's history.

Even a geographic divide between the island's small cities and towns was enough to stoke anger and rash judgments. One of the most vocal activist groups, Friends of Clayoquot Sound, was headquartered in the picturesque town of Tofino, where home prices were double those in nearby Ucluelet, a working-class logging town. Ucluelet's unemployment rate was more than twice that of Tofino, whose work opportunities tended to be more managerial or more traditionally middle-class. Jobs in Ucluelet, by contrast, were industrial or in manufacturing. The two communities came to represent the great divide between worker and environmentalist.

Tensions escalated as the international environmental organization Greenpeace began to expand its involvement in the protests. The organization trucked in outside protesters and paid for anti-logging public relations campaigns. Locals criticized Greenpeace for disrespecting the loggers' perspective, and some Greenpeace members on Vancouver Island eventually became so disillusioned with the group's antagonism that they opted out of protests that blocked people from getting to work. Nelson Keitlah, a leader on the Nuu-chah-nulth council (which was working to halt logging), accused many of the activists of having "literally nothing at stake" in the debate; their efforts, he charged, stymied any cooperation that might happen. (Greenpeace later claimed that Keitlah's nation had been bought off by logging companies.)

Protesters were picked up and physically removed from blockade sites where they had set up camps in the treetops and at access points to logging land. Eventually, more than 900 people were arrested. But the protesters were successful in pressuring the government of British Columbia to conserve 34 percent of forests in the island's Clayoquot Sound region.

All this happened at a time when the broader conception of Pacific Northwest society was rapidly changing. Portland, Seattle, and Vancouver were adopting high-tech economies over shipping, heavy industry, and export businesses. The shift was not only logistical but philosophical and moral. Many people struggled, watching their public reputation flip from useful worker to allegedly amoral force. Timber workers were stuck between corporations—which often treated them poorly—and income. They preferred jobs, not activism.

Unions, challenged with guiding their members through these changes, had chosen to whip up anti-environmentalist anger rather than echo the prewar cautions against corporate overlogging. In reality, though, employment in the timber industry had waned for many reasons, including overharvesting. Timber companies had mechanized forest jobs and were exporting raw timber to Asia for processing. The industry had been experiencing a steady decline in employment for decades; whereas the early 20th century saw logging contribute to employment for 63 percent of Washington workers and 52 percent of those in Oregon, by 1955 tens of thousands of those jobs had already disappeared. By the mid-1990s, only 6 percent of Oregon's population earned its income from logging. Families reported living in tents and camper vans, but they felt tied to the region and resisted moving away from it. Still more felt they were too old to retrain, and many had little more than a tenth-grade education.

Large-scale, clear-cut logging firms were at the center of this change, whether they were harvesting Douglas fir or old-growth redwood. Between 1850 and 1990, ninety-six percent of redwoods disappeared due to logging. In 1985, Houston businessman Charles Hurwitz purchased many acres of Humboldt's old-growth redwoods by acquiring the company Pacific Lumber; in doing so, he came to own most of the region's remaining redwoods. Hurwitz's acquisition was funded with junk bonds, and to quickly earn its investment back, Pacific Lumber doubled its timber harvest, raided pensions, sold off assets, and shifted its focus from selective logging to clear-cutting. One photo shows

Hurwitz and his sons watching a "logging show" in the woods mere weeks after the acquisition, the slopes below them being stripped of redwoods. This bald-faced disregard of a precious dwindling resource shocked environmentalists and inspired mass demonstrations across the region. Rhetoric became heated. Loggers saw protests against clear-cutting as protests against jobs, the dismissal of a way of life at a time when it was on the verge of extinction, and a showcase for misguided romanticism about the forest.

Meanwhile another redwood market had developed, apart from the domestic one that Hurwitz was feeding. Demand was booming in Europe, where redwood burls were being turned into furniture and used in the flashy consoles of luxury cars. Burls—knobbly, pregnant, bubbling protrusions—sprout from a redwood's trunk or the forest floor at its base, entangled in the root system. The largest burls grow underground, near the roots. (The most impressive burl specimen, measuring 41 feet across and weighing 525 tons, was unearthed in 1977.) Inside each burl grows a smooth, knot-free wood, containing much of the DNA a tree species has evolved over thousands of years. The wood is so beautiful that it needn't be stained, only polished.

Contracted by logging firms to harvest any burl that remained beneath their already-logged stands, loggers would carry picks into the forest, digging around the base of tree trunks and keeping an eye out for bulges that humped under the ground, like onions in a garden. Using an excavator, the crew would loosen the tree or stump from the earth until its root system snapped, peeling back with a crack and bringing up the root ball and large burl attached to it.

Burls are often covered with sprouts and buds. When a redwood falls to the ground or is cut from its base, the burl is called upon to seed and shoot new saplings up through the soil, a small part of that redwood reborn. In *Woodlands,* tree researcher and historian Oliver Rackham posits that redwood burls may have evolved to sprout as an adaptation against the grazing of dinosaurs. "But what could the mightiest dinosaur do to these giant trees?" he asks.

After hooking it to a backhoe or excavator, loggers would drag the burl to a landing, where it was cleaned and trimmed with a chain saw. Exporters from Germany and Italy would fly in to examine these burls, remembers one logger, "old, rich men walking along the burls, picking out those they wanted to buy." In the early 1990s, burls sold for about 10 cents a pound, but in some cases as much as $5—a hefty earning for a burl weighing 15,000 pounds. It was not uncommon for a single burl to be sold to a car manufacturer for $450,000, or for a buyer to take home 100 tons at a time. Sometimes, remembers the logger, teams would cut more burls than they were entitled to without telling the landowner, then pocket the profits themselves.

On the other side of the Timber Wars were protesters such as Darryl Cherney, who had fled from Manhattan to Oregon before heading south and landing in Garberville, California. When he arrived in Humboldt County in the late 1980s, he joined the hippie community that had grown angry at what logging had done to the surrounding land. Cherney would become a catalyst of the anti-logging movement in Humboldt, whose members adopted such names as Oak and River and Harmony. The most famous of them, Julia "Butterfly" Hill, lived for years in the canopy of a redwood she had named Luna, on Pacific Lumber property.

Dubbed "tree huggers," these dissenters were deemed sentimental preservationists; within the staid logging communities, however, they were viewed with distrust. Many loggers suspected protesters of having been "brought in" from elsewhere in order to bring everything down. Cherney had told another activist, Greg King, that he could feel the pain of trees as they were cut by saws. Loggers viewed such sentiments as uselessly romantic, but coastal redwoods in particular have evolved in ways that can make them seem otherworldly and spiritual. After being struck by lightning, for example, a redwood may grow a new, "reiterated" trunk high in the sky, which can dwarf its neighbors rooted in the ground. In one notable case, a 140-foot-tall redwood—declared "a freak of nature of unusual interest"—was found to be growing from another redwood's limb. In those same branches, 200 feet above the forest floor, huckleberry bushes grow in cavities, whole ecosystems existing independent of the ground.

Cherney would become active in the Humboldt chapter of one of the most influential environmental activist groups in American history: Earth First! The group's motto—NO COMPROMISE IN DEFENSE OF MOTHER EARTH—and its logo of a green raised fist appeared on bumper stickers throughout the Pacific Northwest as Cherney made his way south. Inspired by Edward Abbey's fictional accounts in *The Monkey Wrench Gang,* Earth First!'s activities in the late 1980s brought the outfit a reputation for dangerous radical activism, including tactics such as tree-spiking—driving large nails into the trunks of trees, with the goal of destroying the logging equipment used to harvest them. Tree-spiking threatened the lives of loggers; machinery sometimes broke as it came into contact with the nails, shooting fragments of sharp metal into the air. By the 1990s, Earth First! had become a target of FBI surveillance, though members of the Humboldt chapter had renounced tree-spiking.

In Humboldt, the Earth First!ers focused primarily on a 3,000-acre tract of land called Headwaters Forest, owned by new Pacific Lumber chief Charles Hurwitz. But they also set up camp on other slices of private land known to harbor old-growth, trespassing in camouflage and wearing branches to blend in. By the late 1980s, the atmosphere in Humboldt was tense, and it was easy for anyone who earned money from the logging industry to conflate activists with park rangers, biologists with protesters.

Activism provided a visible scapegoat for many people upset about the declining economic viability of logging. Halts on logging presented an abrupt example of what had been happening gradually: one day, hundreds of people suddenly had no work to go to. Loggers found themselves at a crossroads, and in many cases their reactions appeared reactionary and inappropriate. Bumper stickers and signs were found all over: SAVE A LOGGER, EAT AN OWL or EARTH FIRST, WE'LL LOG THE OTHER PLANETS LATER. One sticker distilled the animosity perfectly: ARE YOU AN ENVIRONMENTALIST, OR DO YOU WORK FOR A LIVING?

Loggers were frustrated with media depictions of the crisis: editorial cartoons portrayed loggers sitting on stumps, waiting for small seedlings to grow large enough to cut down. One satirical drawing placed a Leatherface mask on a man with a chain saw and was captioned *Oregon Chainsaw Massacre*. These were stereotypes of working-class loggers rather than critiques of logging companies, and they had the effect of pushing loggers further away from conservation. In 1990 Robert Lee, a sociologist at the University of Washington, told newspaper columnist Jim Petersen that there would eventually be "human consequences for the victims [of stereotyping]," including cynicism and depression in a cycle difficult to escape. Lee argued that the breakdown of community would lead to personal and family problems, including substance abuse and divorce.

On the ground, there were incursions on both sides. Candy Boak, the wife of a logger, went undercover at activist meetings and sabotaged Darryl Cherney's first attempt at a tree-sit. A protester approached a logger in the woods, grabbed his ax, and threw it down a ravine; he responded by punching her. Later, Cherney left a tree-sit in handcuffs and told a crowd outside the county courthouse that he was a "prisoner of war in the fight to save the redwoods." He began working consistently with a former union organizer named Judi Bari, who before moving to Humboldt had studied at the University of Maryland, where she claimed to have majored in "anti-Vietnam rioting."

Having found work in California as a carpenter, Bari saw her adversaries as Louisiana-Pacific, Georgia-Pacific, and then Hurwitz, not loggers. She was open to discussion and appeared on local radio shows, listening as loggers called in to tell her about their lives. "I mean, logging was my life," a man named Ernie told her. "It's a tradition. It had always been happening, and always before it looked like there was always going to be enough trees." Bari saw her role as messenger to timber workers, though it's questionable whether she achieved that goal. Her writing is infused with frustration: "[B]y and large, timber workers are either doing the companies' dirty work, or keeping their mouths shut." She wished that the companies operating in the area hadn't harnessed the fear of instability so easily, yet she also criticized the environmental movement for its lack of class consciousness.

The movement was filled with activists like Cherney, who came from elsewhere and involved themselves in communities where they had only shallow roots, if any at all. They favored violent rhetoric: timber companies were "raping" the forest, for instance. Of this, Bari was hardly innocent—she branded Harry Merlo, the former CEO of Louisiana-Pacific, as "the ultimate tree Nazi." She called loggers who disagreed with activism "the equivalent of the white racists in Mississippi.... They're being used by the system. But they are people who are not real bright who have bought into it." In one speech at an Earth First! demonstration, Bari belied her supposed sympathy for the local community of working-class loggers: "There's too much inbreeding too," she said. "It's a rural area. The genetic pool is not large, and some of these families have lived here for five generations."

Loggers and mill workers were incensed by language like this—the forests they were accused of raping were the same precincts in which they went hiking and camping, the same places where they had built their homes. They dreaded a future without timber to harvest, and that angst morphed easily into anger: "You fucking commie hippies, I'll kill you all!" Bari remembered someone screaming at an Earth First! roadside blockade. When an Earth First!er chained himself to a logging truck at a red light in February of 1990, the driver of the truck later told the *San Francisco Examiner,* "I think I own this land."

The violence in the region was reaching a boiling point. Bari and Cherney were working to make the summer of 1990 "Redwood Summer"—an environment-focused reboot of the Freedom Summer for civil rights in the 1960s. The summer featured activists dressed as owls, tree-sits in the forest, and a festival along the Eel River. At the same time, logger John Guffie told contractors to "leave their hotheads at home, because if they touch one of these guys out there chaining themselves to the equipment and gates, it would be a big lawsuit." At a pro-logging protest at a chip mill, one woman held aloft a troubling sign: IF YOU TAKE MY HUSBAND'S JOB, HE TAKES IT OUT ON ME.

Resentment toward logging restrictions, activism, and government oversight lodged itself deep in the psyche of loggers during the Timber Wars. There was deep financial need on both the individual and community levels. By the mid-1990s, Forest Service rangers had become attuned to the choppy fluttering of a single chain saw whirring to life in the deep woods. Across the country, timber theft was on the rise.

The Forest Service founded a special Timber Theft Task Force in 1991, which was sent into the woods to cruise for stumps and monitor high-value stands. For three years, task-force members conducted stump cruises among timber stands designated for recreational use. They monitored valuable plots of forest and ran investigations into thieves who had until then gotten away with poaching. Many of these thieves were white collar, corporate entities, logging beyond boundaries and passing the contraband timber through mills.

The introduction of the investigations branch coincided with shifts toward policing and heavier law enforcement in both the Forest Service and the Park Service. In the wake of the 1990 slaying of a park ranger in Gulf Islands National Seashore in Mississippi, as well as high-profile cases of armed battles and drug smuggling through national parks, rangers were required to become professional law-enforcement officers, and they were sent for police training.

The timber-theft unit left footprints embedded in the soft soil of the Pacific Northwest. But investigators were also present in the stick-straight stands of Pennsylvania and Vermont, and in the state forests of Ohio, New York, and Wisconsin. The white oak, black walnut, and maple trees that stand sentinel on the East Coast were valuable for other means, but they were every bit as in demand as those impressive redwoods in the West.

The Forest Service—which relied on the threat of litigation and large fines to deter forest crime—found it difficult to dissuade poaching. Each day, rangers would note the coordinates of stumps and felled logs on maps and in notes. Publicly they struggled against the assumption that poaching timber was a crime easy to get away with, but in fact that assumption was true: it *was* simple, and that's why so many did it. The Timber Theft Task Force was pursuing not just small-time poachers but big-name corporations—among them Weyerhaeuser, suspected of illegal logging in an Oregon national forest.

Only four years after its founding, the Forest Service disbanded the special task force. The decision to do so was shrouded in secrecy and conspiracy theories: might Weyerhaeuser have lobbied the White House to shutter the special branch, as some environmentalists believed? The unit had also become unpopular among the ranks of Forest Service and National Park Service rangers, who saw the investigators as stepping on their toes and providing a near-paranoiac degree of oversight. For its part, the Forest Service insisted that Timber Theft Task Force rangers were simply being reassigned to work at the local level across the country.

Eventually, the federal government was called in to try to broker a détente in the Pacific Northwest—both rhetorical and in the forest.

During his 1993 presidential campaign, Bill Clinton promised to solve the issue of discord in the Pacific Northwest. After his inauguration, he arranged an April 1994 summit in Portland. At the summit, Clinton, Vice President Al Gore, and their top decision makers sat at a long, wooden boardroom table that had been placed on a stage surrounded by stadium seating in a conference center. Every seat was filled by someone with a vested interest in the Pacific Northwest timber industry: community leaders, politicians, logging executives, loggers, priests, teachers, and biologists from across the region. All had come to testify to the challenges that a shuttered logging industry would create in their lives, and for the world around them.

Some of America's foremost biologists urged the table to consider the forests with "caution and humility." Forests, said biologist Jerry Franklin, sitting across from President Clinton, are "more complex than our wildest imagination." In the coming days, these experts would outline what continued harvest levels would mean: scarce forests, 480 species at risk. Also sitting at the table were members and colleagues of "the God Squad"—the Endangered Species Integrity Committee, which had the power to add, and to make exceptions, to the 1973 Endangered Species Act, essentially playing God with the fate of various species.

The biologists sat alongside historians and social scientists, who outlined the powerful force that logging had played in the region's history and identity. Vice President Gore had opened the summit by acknowledging that logging was woven into the nation's cultural heritage. Later, a logger explained how logging had been in his family for 200 years. A mill owner from the town of Forks, Washington, said that his "American dream had turned into a nightmare," one filled "with blood and gore." Louise Fortmann, a poverty researcher from the University of California Berkeley, explained why outside forces such as government and urban-based environmental groups "angered" timber communities: "[These organizations] are not impacted by decisions, don't have family ties, and see jobs as uniforms," she said.

Northern California's timber workforce was represented by Nadine Bailey, a mother and logger's wife whose husband had been thrown out of work as logging restrictions tightened. "We need a solution that involves local people," she said. "Don't send us money . . . we need to work. We need that pride."

But the most emotional testimony came from Seattle's archbishop, Thomas Murphy, who had traveled the roads of Olympic

National Park, talking with locals and spending time in timber towns throughout the peninsula. His was a message of "lost home," he said. "Do you know what it's like to work 20 years, then sleep in a pickup truck?" he asked the table. "A way of life is dying."

PART II

TRUNK

Chapter 6

THE GATEWAY
TO THE REDWOODS

"There is nothing there . . . The town is dead."
—Danny Garcia

The Redwood Highway is a stretch of California's Highway 101, an artery that skirts the Pacific Coast and stretches from Los Angeles up through the state's northernmost counties and into the heart of Cascadia. If you are driving north from Humboldt County's largest city, Eureka, your passage is guided by the unfurling crystal waves and white-sand beaches of Big Lagoon and Freshwater Bay. In fact, despite the iconic, glittering shores to the south, Humboldt's 110 miles of coastal, craggy Pacific Ocean form the largest unbroken sand line in California. When you make the trip, it feels as if a curtain is parting, unveiling immaculate vistas between cliffs and sea.

Orick sits along a slight bend in the highway, its businesses and houses stretched long and thin. No longer technically a town but a "designated census place," Orick is home to just under 400 people. ("And I'm quite certain that counts 80 sheep," clarifies Jim Hagood, owner of the town hardware store and the son of pro-park advocate and motel owner Jean Hagood.) Its demographics are overwhelmingly White, primarily English-speaking, and mostly over the age of 45. Burl shops, the sole remaining industry, dot the highway, beckoning tourists to visit with intricately carved figures and tables hewn from redwoods. At the peak of the burl industry in the 1970s, about a dozen burl shops lined this stretch of highway leading into the boundaries of Redwood National and State Parks. Now there are fewer than five. (When I checked in 2021, two had just shut down.)

North America's remaining redwood habitat is only 35 miles wide, a narrow belt running along the California Coast Ranges, and it represents a ribbon of some of the Earth's most ancient ecology: just two acres of this land can host nearly 10,000 cubic meters of biomass. The stands of trees that grow north of here, from Oregon to British Columbia, are generally more diverse: western red cedar, maples, yellow cedar, Douglas fir. The four latter species are some of the tallest trees in the world—their trunks can reach 100 meters—and though not as hardy against flooding as redwoods, they grow fast and produce fine wood that is light and valuable.

Orick is the gateway to these forests, particularly the old-growth redwoods protected by the National Park Service and California State Parks, which work together to manage the woods surrounding the town. It is home to the parks' South Operations Center, or SOC (pronounced "sock"). Within its boundaries, the park holds 45 percent of the world's remaining old-growth coast redwood forest, and (as far as we know) the planet's tallest tree. Of the two million acres of coast redwoods that once carpeted the region, just four percent (along a 450-mile-long stretch) remains, most of it along Highway 101.

When Lynne Netz first arrived in Humboldt County in 1993, she worked on commission selling tickets for the Greyhound bus line. Netz and her family spent their free time driving the winding roads that passed through mill towns such as McKin-leyville and Trinidad. Sometimes they went horseback riding in Redwood National Park, skirting the edge of Orick, with the Pacific Ocean lapping craggy rocks to their west. Occasionally whales hugged this shoreline, feeding on the salmon that migrate into rock pools only steps from cool, shaded forests.

The oldest coastal redwood to have its rings counted was 2,200 years old. A bit of its stump, which was growing when Hannibal took his elephants over the Alps, is preserved in Richardson Grove. But trees just as old—already ancient when philosophers in Greece and Rome dubbed them *hulae* and *materia,* or the matter of life—still fill Humboldt's forests. Indeed, redwood trees left undisturbed are virtually immortal: when fire touches a redwood trunk, its bark uses the chemical compound tannin to shield the tree from the flames. Some redwood bark, fluted in long, deep crevices that splinter and meander off, has been measured at two feet thick. Redwoods owe their longevity to their ability to sprout new trees from the trunks and roots of older specimens—making them not so different, really, from human children and parents. "It is nearly impossible to say where a Redwood life ends," writes biologist Donald Culross Peattie in *A Natural History of North American Trees.* "Rather it changes direction and grows on."

The redwoods are the backbone of what entices visitors to the region. One stop for tourists is at a redwood called simply *Big Tree,* located not far from May Creek. It stands 286 feet tall in the middle of a grove, near a guidepost bristling with colorful pointing signs:

THIS WAY TO MORE BIG TREES!

ANOTHER BIG TREE!

EVEN BIGGER TREES!

Somewhere in this forest is a 309-foot-tall redwood dubbed the *Big Kahuna.* Discovered in 2014, it has a base diameter of 40 feet and is estimated to be 4,000 years old. Its size puts it in contention for largest tree in the world, rivaling southern California's *General Sherman* sequoia, as well as *Hyperion,* located in the depths of Redwood National and State Parks. Just over 38,000 acres of old-growth forest are protected within its boundaries, and some of the largest trees—like *Hyperion*— are part of a group of redwoods that have been measured by researchers who keep their location secret.

Until relatively recently, one or two burls might be poached from the park land surrounding town every year. Most often they ended up forged into bowls, crafted into statues, or sold as slabs at the burl shops lining the highway. But in the early 2010s, that changed: so much wood poaching was taking place that Redwoods rangers began calling it a "crisis." From 2012 to 2014, nearly 90 burls were poached from 24 trees. One redwood had been felled in order to harvest a burl growing high up its trunk. The Lady Bird Johnson Grove Trail had been, in the words of one ranger, "all hacked up," as had trunks along the popular Tall Trees Trail. "There was this huge [realization of] *Holy smokes, they're just hitting everything, whenever,* with no thought of getting caught," says Brett Silver, who headed the region's state parks at the time.

A burl serves as both stress response and safeguard of genetic immortality for a redwood. It forms like a callus after trauma or emergency—most likely fire, flood, or high winds. Trunk burls tend to emerge right above the spot where the tree's "wound" hit hardest, growing out and down over the bark as a kind of protective bandage. If the damage is extensive enough, new bark will protrude two feet or more horizontally before growing down to encase the injured area. Shoots and buds sprout from the wood inside, reaching toward the ground—where they will begin to plant roots.

There are some instances where forests, after being clear-cut, have grown anew thanks to the sprouts arising from ligno-tubers. When a burl is cracked open by fire, for example, the old-growth's genes are spread across the ground in much the same way a jack pine's cone cracks open to release seeds. In this way, burls and stumps are the guardians of redwood lineage, ensuring that a tree's evolution will continue despite environmental—or industrial—devastation. "It's as if you took a chunk of the forest floor and suspended it into the air," redwood-canopy expert Stephen Sillett told the *New York Times* in 2014.

Some of the most dedicated redwood researchers in the world say that the true impact of burl poaching is unknown—no one will be around long enough to understand and monitor its long-term effects. But we do know what the burl holds, so we can speculate on what its removal might bring.

Biologists say that removing a burl leaves an opening for disease and infection to attack a tree. If enough is taken, the tree can become "girdled"—unable to grow more rings, its growth permanently stunted. When the part of the tree that facilitates seedlings and new growth is removed, damage occurs not only at the instant of the crime but in the future, when the redwood is affected by other forces (invasive species, drought, forest fire). Old stumps have been found to support saplings, the redwood thus contributing to its community even after its body is gone.

Poaching "dead-and-down" trees, too, has an ecological effect. Dead-standing trees continue to provide shelter for birds and other animals, and dead trees that have fallen are burrowed into by beetles and their larvae—both favorite foods of birds. Trees that have fallen can take hundreds of years to decompose, during which time they supply nutrients to the soil and habitat for animals and fungi. Like burls, new seedlings take root in downed trees. In these ways the body of the tree furnishes as much as it can until it completely decomposes.

Apart from burls, there is much about trees that we are only now beginning to understand. At times they can feel like unknowable, practically unbelievable specimens of ecological evolution. Below our feet stretches "the wood-wide web," vast underground communication networks that spread information to ensure every tree works well with one another. The network is practical—it allows resources to be shared wisely, sending alerts when one tree has been attacked or depleted and cueing healthy trees to support it.

Trees also communicate through an olfactory language: they can detect attacks from insects and animals, for instance, and pump out a scent through their leaves to dissuade further visits. Sometimes a tree will warn its neighbors that a bug is munching on its leaves—trees can recognize dangerous saliva—by sending a message through the wood-wide web to other trees, which then secrete that scent to ward off any pests. Or their tannins may strengthen, making their bark or leaves unpalatable or even lethal.

There is not a forest in the world that can be completely known to us: they are immensely surprising, continually refreshed. Forests exist at the intersection of root, trunk, branch; moss, fungi, stream, bird. Sometimes you can see whispers of a forest ecosystem in the eroded sections of earth along roadsides or hills that have been dredged for other projects. There are impressive examples of this in Redwood National Park: huge stumps and curving, switchback roots that appear as if carved into the land, merging with one another. Scientists have stumbled on the remains of ancient woods in this way, locating root systems that continue to feed the forest long after the body of the tree has disappeared. In this sense the tree's influence extends beyond the scope of its body; it remains an ancestor. The trees that once towered here on Yurok land continue to inform the actions and reactions of trees in front of us today.

Still, the notion of burls as *rare* and valuable is relatively new. The head of the Humboldt County Historical Society when I visit, Jim Garrison, tells me that his grandfather had a "burl tree," from which he would cut off pieces and send to cousins else-where. Ron Barlow remembers cutting burls off trees near Orick as a child and selling them at tourist markets in Eureka. In the 1980s, a resurgence of interest in burl items such as sculptures brought shoppers flocking. Nowadays, by contrast, interest has slowed—as Garrison says, the "well has gone dry."

Despite this, rangers at the SOC found themselves confront-ing a crime spree that was nearly invisible: one that takes place in the lushest of woods, surrounded by towering old-growth, in the dead of night. To track down the perpetrators, they created a map pinpointing a constellation of poaching sites around the town of Orick: eight locations in total, all just steps from the Redwood Highway—and some off the region's most popular hiking trails.

The underside of the thick evergreen redwood sorrel that grows on the forest floor beneath redwood trees is a deep, dark purple. Often when he is searching the forest for signs of black bears, biologist Preston Taylor keeps his eyes open not for the black smudge of an ambling bear, but for the color purple: the grape-stained evidence of where a bear may have foraged, trampling the sorrel with its paws and leaving a streaked path in its wake.

On April 19, 2013, Taylor, then a student researcher at Humboldt State University (HSU), entered Redwood National and State Parks to collect data, hoping to perhaps also spot a black bear in its habitat. Taylor was nearing the end of his undergraduate program in wildlife management at HSU, where he was researching scent marks left by black bears in the forest. He spent many of his days in the woods looking for "rub trees"—trunks against whose bark the bears rub themselves to signal potential mates during breeding season. Eyes on the ground,

Taylor parsed indents in foliage for bear tracks and purple blotches—which would lead him, he hoped, to a rub tree.

That spring day Taylor set off on one of his customary hikes, following a trail beside Redwood Creek, a 62-mile-long river that originates in the Coast Ranges and cuts down through a deep forest, flowing northwest and feeding tributaries before it enters the boundaries of Redwoods. There the creek weaves its way through several groves of redwoods before winding through farmland and emptying into the Pacific on the outskirts of Orick.

About half a mile up the path along the river basin, Taylor caught a flicker of purple in the corner of his eye. Some of the sorrel had been disturbed by prints of some kind, with a makeshift path leading into the forest and up a hill.

Taylor knew the area was home to some of the world's largest redwood trees, so he eagerly followed what he assumed was the black-bear trail for about 200 yards into those stands. The trace eventually petered out, leaving Taylor standing in a deep thicket of brush. Gaining his bearings, it took him a minute to focus on the massive redwood reaching up into the sky before him.

That's when he noticed the gaping hole in its trunk. A rectangle more than eight feet high had been sliced from the base of the tree, with unnatural edges that only a chain-saw blade could have made. The trunk looked like it had been roughly peeled, one side of it a faded taupe compared with the deep brown of the remaining bark. Taylor walked close enough to peer into the heartwood of the tree. Around the tree stood smaller chunks of wood, which had been cut into arm-sized slices.

Taylor stepped back, realizing that he had discovered a poaching site. Taking in the full extent of the redwood, he craned his neck upward, tilting his head toward the sky, following the length of the ancient tree.

Chapter 7
TREE TROUBLES

"He was pretty much born into it. That was his life."
　　　　　　　　　　　　　　—Cherish Guffie

As Redwood National Park was expanding its boundaries in the 1970s, logger John Guffie's son, Chris, had begun his own process of lumbering osmosis—he would be 19 and ready to enter the workforce at the time of the park's expansion. By then he had already earned a nickname from an old-timer: "the Hammer."

"I've been setting chokers since I was 13," he says. "I'm a logger by trade. I've been doing it ever since I was just a pup." Chris, it was assumed, would one day join his father's company, Guffie Timber Cutters, and he had been spending his summers helping in the woods, at logging camps. "I was always taught I have to work for what I get," says Chris. "If I can use these two hands, I can get what I want in life."

Chris Guffie was friends with Terry Cook, who lived in a house at the very southern edge of town. In 1970, Cook's family had wound its way north along Highway 101, traveling from

Tennessee to the very tip of northern California. His parents were moving the family for work, and they stopped in Orick while Terry's dad, Harriel Edward Cook, looked for work in the woods; his mother, Thelma, had a friend working at the Palm Cafe and Motel. The family bought a small house facing the Redwood Highway; it had been lofted up on stilts, a response to the 1964 flood that had overflowed nearby Redwood Creek, which skittered along the house's lush backyard. The Cook family would eventually grow to include 11 children, who often gathered firewood from the beach to feed the woodstove.

In 1971, Harriel died when his car hit a panel truck on a bridge in the middle of Orick. Four years later, one of Terry's older brothers died after being hit by a logging truck while riding his motorcycle. And then a few years after that, another brother, Timmy, was paralyzed in a motorcycle accident. Thelma, stricken with grief, was left to raise the family alone. Her older children went out, looking for work.

In Orick, that meant working with wood. The family's boys went to work at Arcata Redwood, where Harriel had been employed. They were hard workers and earned a reputation for being keen to help their neighbors. "We'd call Mother Cook up when we needed crew in the hay fields," says Ron Barlow. "And we got the best. They were strong boys." Terry eventually moved back into the house on stilts. One of his sisters, Charlotte, had moved south to Los Angeles and had a son named Danny.

Terry Cook was a witness to the slow devastation of not just his family but the town's logging industry. He watched his neighbors head off on the Peanut Convoy and return angry. He was 17 when the park expanded, and he says he applied for a maintenance job but was never hired. His jobs in the woods and at mills—running an edger and pulling green chain—were inconsistent.

Chris Guffie never got a chance to take over the family business, which John shut down in 1980. Chris noted the high-pitched fervor of the community debate about the expansion of Redwood National Park. He listened to the rhetoric: the park would mean "annihilation" for the town. "They want to control everything," his dad said. John called the park officials "parasites."

Charlotte Cook's son, Danny Garcia, soon began spending his summers traveling north from his family home in southern California, meandering through Sacramento and the lush valleys of Mendocino County into Humboldt to visit the Cooks. When Garcia was nine years old, Charlotte took her own life. He and his older sister were raised by their father, who drove gasoline trucks for Union 76 and Arco gas stations. The siblings went to Orick for two weeks every summer, then south to Los Angeles for another two weeks to visit their paternal grandparents.

But it was Orick that gave Garcia a home. Sometimes an uncle would pick him up in Sacramento and they would drive for hours on Highway 101, gazing out the windows at the massive logs loaded on the backs of trucks. He spent days running around the forest that surrounded Thelma's house. "It was beautiful up here," he remembers. As a boy, Garcia was enchanted not only by the forest but by the freedom it afforded and the land on which the tall trees grew. His uncles would let him spend all day in the woods on his own. "Get my hands dirty," he says now. "Nobody gave a shit."

After leaving high school in grade 11, Garcia moved to Orick permanently. At 18 he moved in with Thelma Cook, who continued to nurse his uncle Timmy after his motorcycle accident. He helped out around the house—picking up food and doing yard work so Thelma could care for her son. And he spent time with his uncles. "Terry's got a big heart for people," Garcia says. "[His] physical roots are so deep in that town."

By hanging around the Cooks' house and accompanying them on trips to harvest firewood, Garcia watched and learned how to start and use a chain saw. He had entered into a pattern familiar to the men around him. He had begun the osmosis that John Guffie talked about: how to properly fell a tree, guiding its fall in the right direction, "bucking" it up into small chunks for transport in the back of a truck. "He didn't know the first thing about it until he worked with me," Chris Guffie says today.

Over the years, however, Garcia began to feel the walls closing in on him in Orick. He likens the sensation to having your car break down in the middle of nowhere—you've no cash to fix it and no way out. He had entered the town's orbit, was "running around, causing trouble, being a kid." But eventually Danny Garcia felt that he had to escape.

Late in 1993 he moved north to Washington's Jefferson County, the boundaries of which butt up against Olympic National Park, the Hoh Rain Forest, and Olympic National Forest. Chris Guffie was already there and had leased a salvage sale—a parcel of dead-and-down timber land sold by the state because the wood is damaged or infected, open to the public to buy for harvest—not far from the town of Forks. Guffie set up a small shake-and-shingle mill in which he planned to process the wood that he gathered from the salvage site, and he brought Garcia in to help with the work. The cedar trees along the parcel boundary were marked with red paint, clearly visible even in thick Northwest fog.

"That's where my tree troubles started," says Garcia, who says this was the first time he poached a standing tree. "I asked [Guffie] why we were scavenging wood: *Let's just go find a tree and cut the damn thing down.* I saw the sparkle in his eye when I said that."

The boundary lines did little to deter Guffie or Garcia, who slowly began felling standing cedar beyond them. The wood they took was old-growth, protected by the Endangered Species Act, and home to northern spotted owls. The pair had forged a brush trail in the forest where there previously was none, and they bucked the wood into blocks to transport by hand.

Into the spring and summer of 1994 the two sold the wood to a mill, which turned it into guitar-body blanks that were then sold to instrument manufacturers. Sometimes the cedar was sold to local artisans to make arrows for bow-and-arrow sets. The payoff in crossing the line was well worth it: had they stuck to harvesting wood from the salvage sale, they would have earned about $600 per cord. Beyond those boundaries, by contrast, they were able to net about $2,000 per cord.

As they went, sometimes they left long intact trunks behind. One day Guffie asked a local helicopter pilot to help him remove them and gave him the coordinates. Thinking the haul sounded suspicious, the pilot alerted the Washington State Department of Natural Resources.

Visiting the salvage site some weeks later, an investigator spotted boot imprints on the makeshift trail. Strewn on the ground nearby were a gallon jug of chain-saw oil, a Snickers wrapper, a package of sunflower seeds, and empty cans of Pepsi and beer. The cans sat near three large cedar stumps outside the approved cutline, the stumps covered with forest debris in a bid to conceal them. When timber cruisers examined the spot later on, they estimated that 20,000 board feet of timber had been poached—a value of more than $33,000. Access points and trails had been carved through the woods, leading to unofficial backroads not plotted on any map.

After the investigator's visit that day, he stopped by an unrelated salvage site and found a man named Robert Jackson there, cutting wood. Jackson's boots appeared to be the same size as the imprints on the trail. Peering into the cab of Jackson's truck, the investigator spied sunflower seeds and a beer can between the seats. When the investigator visited him at home later that day, Jackson admitted to poaching wood with Chris Guffie and Danny Garcia. His admission was corroborated by a neighbor, who reported seeing Garcia show up with loads of wood in the middle of the night.

In autumn 1994, Guffie and Garcia were charged with poaching from state land in Washington. Prosecutors for the Natural Resources Department felt the pair were unlikely to be able to pay the full value of the poached timber, or even the joint restitution amount of $16,975. Both men pleaded guilty to the theft, and both were sentenced to 30 days in jail.

But they also both vanished.

Their sojourn on the Olympic Peninsula had lasted less than a year. Rumors circulated that Garcia had returned to Orick. As for where Guffie had fled, no one knew.

Chapter 8
MUSIC WOOD

"I've been at it for so doggone long. It's like Yogi
Bear and the park ranger."   —Chris Guffie

The rectangular gash that Preston Taylor stumbled upon near
Redwood Creek in 2013 dug about halfway into the trunk
of the tree—close to two feet deep. He could tell that the slice
was fresh: not only was dry sawdust scattered around the base of
the tree, but the wood itself was a bright, pale brown. Bark slabs
and wood chunks had been left behind on the forest floor, including one about the size of a couch.

Redwoods contain chemicals called terpenes that release an
earthy, musty smell. Moisture intensifies this distinctive aroma,
and because the forest was wet that day, the air was thick with
fragrance. Circling the damaged tree, Taylor examined the cut
more closely. He saw that the heartwood had been exposed—a
sure sign that the cut had been so deep it could compromise the
tree's ability to stand.

Taylor hiked back down the trail, climbed into his vehicle,
and drove directly to the South Operations Center in nearby

Orick. Arriving in the late afternoon, he told the front desk clerk that he needed to report a crime on park land. But then Taylor quickly grew nervous about discussing what he had seen, so he was ushered into a private office. Taylor took a seat opposite a park ranger and relayed everything he had discovered, including the poaching site's GPS coordinates. He agreed to meet a ranger at the trailhead the following day and guide her to the scene.

As planned, Taylor met National Park Service ranger Rosie White at the trailhead the next morning, and the two began walking up the trail along Redwood Creek toward the poaching site. Though White was armed, Taylor could tell she was nervous. Might they interrupt a poacher along the way, inciting an attack? What if someone tried to rob them of the camera White carried?

When they got to the site, Taylor stopped in his tracks. "This looks different than it did yesterday," he said. Some of the burl chunks he had found the day before were nowhere to be seen. Not only had he stumbled across a poaching site, Taylor concluded, but he had apparently done so while the thieves were in the very act of removing the haul. He watched as White began to photograph the crime scene and measure the scarred redwood and its remaining burls. Before they left the site, White cached some motion-triggered cameras in the surrounding foliage, hoping to capture images of the poachers if they returned. When she visited the site again five days later, however, nothing else had been removed, and none of her hidden cameras had been activated.

Investigating timber theft poses many challenges, starting with the uniqueness of its setting. Whereas a theft from an urban area leaves investigators with the option to examine any evidence left behind, in a forest that evidence—think sawdust or evergreen needles or deciduous leaves—is easily degraded or simply blown away. Then there's the physical danger: a park ranger or law-enforcement officer alone in the woods makes a vulnerable target. Precisely for this reason, most rangers focus instead on stopping and searching vehicles for suspected purloined wood as they travel local roads and highways.

Back at SOC headquarters in Orick, a handful of rangers began trying to reconstruct the pathway that the stolen redwood must have traveled on its way from forest to market. They began by visiting local burl shops, their yards overflowing with the gnarled, knotty, bulbous excesses that grow on trees.

The surge of poaching that Redwood National and State Parks was experiencing around this time was not unique to northern California. Theft was a constant forest threat throughout the entire Pacific Northwest, including in Garcia and Guffie's former community of Forks.

A few months before Taylor stumbled upon the Redwood Creek poaching site in 2013, US Forest Service special agent Anne Minden, who had investigated forest crimes in the Olympic Peninsula for 20 years, told the *Seattle Times* that timber poachers there had "completely trashed the national forest." One case in particular—that of Reid Johnston—would become a prime example of the poaching that was sweeping the region's woods.

Brinnon, Washington, a town with a population of just over 800 on the other side of the Olympic Peninsula from Forks, runs right along the border of Olympic National Park and the Olympic National Forest. Stands of Douglas fir, some of the most productive trees in the world, fill the Olympic Peninsula, stretching from the Skeena River in British Columbia south through the Sierra Nevada. A Douglas fir grows arrow-straight; second only to the redwood in terms of height, it is an iconic conifer of the

Pacific Northwest. Although Douglas firs are not as imposing as redwoods, they grow in impressive, noble stands—layered and thick. They grow swiftly and intensely, and their growing season is long. Most plywood comes from Douglas firs, and they have tremendous regenerative power and will continually repopulate themselves.

Reid Johnston's prominent family had lived all over Washington state before settling down in Brinnon in the 1980s. Stan Johnston was known as the town's unofficial mayor until he died in a car accident in 2011, when his truck spun off the road and struck a tree. A year later, Stan's middle son, 41-year-old Reid, was sentenced to a year in jail and slapped with an $84,000 fine for stealing 102 trees from a chunk of the Olympic National Forest that butts up against his parents' property.

Reid Johnston was a logger by trade, a new father, a forestry-degree dropout, and a small business owner. He was also a suspect in four other timber-theft cases, as well as a rumored methamphetamine addict. He harvested high-value maple and cedar trees from the area for his company, Sound Maple, which then sold the wood to instrument manufacturers.

So-called "music wood" of the type Johnston was selling comes from "figured" maple—wood with a distinct grain that is enhanced through milling, revealing striking patterns and rivulets through the lumber. It generally appears in one of two patterns: The caramel-toned wood of *flamed maple* displays bold tiger stripes, whereas *quilt maple* boasts the appearance of ripples radiating across a smooth, glassy lake. Music wood is exceedingly rare, and it commands premium market prices: it often sells for 100 times more than non-figured maple. Products incorporating it have been described as the "closest you can get to hearing an entire orchestra emanate from one instrument."

This natural beauty helps explain why the maples poached in

Washington, as well as the Sitka spruces that grow in Alaska's Tongass National Forest, make highly sought-after soundboards—the front-facing parts of acoustic guitars. (Indeed, poaching and deforestation of the Sitka spruce in Alaska were once so common that Greenpeace flew executives from the world's top guitar manufacturers to the Tongass just to show them what was being destroyed for instruments.)

Johnston first attracted the attention of Forest Service rangers thanks to a 300-year-old Douglas fir. The tree soared 155 feet into the air and measured eight feet in diameter. When Johnston presented a few stolen sections of its trunk at a lumberyard in Shelton, about an hour's drive from Brinnon, the owner judged this nice Douglas fir perhaps a little *too* nice and reported it to the Forest Service. The rangers started their investigation on the ground, conducting interviews with the mill owner before moving on to Brinnon. They quickly began to piece together their case against Johnston.

It would become one of the largest timber-theft prosecutions in the state's history.

The Forest Service rangers suspected Johnston of poaching trees from the forest behind his family's property, in an area known as Rocky Brook, which runs along the eastern edge of the Olympic National Forest. Rocky Brook had been forged from a forest fire 140 years earlier, and its stands of western red cedar, Douglas fir, and western hemlock are categorized as "mature growth" (sometimes called "pre-old growth"). Western red cedar, the same species poached from the Carmanah Walbran in British Columbia, is also known as the "canoe cedar" because its trunk is readily hollowed out for navigating waterways. As is the case with redwoods, only a few red cedars remain; the largest of them, with astonishing circumferences over 62 feet, are in Washington's Olympic National Park and Forest. But with western red cedar paneling our homes and making up more than 80 percent of roofing shingles in the US market, we actually live within it and beneath it.

The poached wood that Johnston was shopping around came from trees that had survived that 1880s forest fire and kept on growing. They furnished critical habitat for the marbled murrelet and the northern spotted owl. Left untouched, they would have grown for another 700 years, becoming the old-growth of tomorrow.

The rangers made a trip out to Rocky Brook. When they arrived, they found that trees just a few dozen meters beyond the private-property line had been cut or damaged. (Maple trees are commonly damaged but not harvested by potential poachers, who check their value by using an ax to cut off a piece of bark to expose the grain and confirm it as quality music wood.) "Private citizens often encroach on national forests next to their land because it gives them a solid defense," says Matthew Diggs, a state's attorney who was assigned to prosecute the case. Johnston's older brother, Wade, told rangers questioning witnesses that he had been at the site of the Douglas fir with his brother; upon noticing the Forest Service boundary marker, however, Wade had told Reid to leave the tree alone. He then apparently left the scene. Ignoring his brother's warning, Reid went ahead with his scheme: he moved the boundary markers some 75 feet away, then planted them back in the ground—facing the wrong direction.

In interviews, the rangers report speaking to people who said they were involved in hauling the fir—and other trees—to the lumberyard in Shelton. The investigation linked Johnston to the thefts of another 99 trees—50 western red cedars, four Douglas firs, and 45 big-leaf maples. While the Douglas fir is an emblematic tree for Washington (poachers don't need to check the grain within because a fir's value lies in its height and diameter), the maples, often used in manufacturing cellos and guitars, are much more valuable: a specimen with an especially beautiful grain can fetch more than $10,000.

The statements from locals, paired with the evidence just outside the Johnston family property, secured law-enforcement officers a search warrant for Reid's home. Upon entering the residence, they found papers advertising trees as music wood, which Reid had posted for sale online. They also found correspondence proving he had been trying to sell the Douglas fir to an exporter, who would have shipped it to Hong Kong.

Reid Johnston was charged with theft of and damage to government property. Though he denied having moved the boundary markers, the investigation at the site provided a strong piece of evidence: the Forest Service land had been logged decades before, leaving a cutline that made it obvious where the Johnston property ended and the old-growth forest began.

Johnston eventually accepted a plea deal in the case. He was sentenced to a year in prison and $84,000 in damages. But that fine was far less than the value of the poached stand, which a 2011 ecological and economic assessment estimated at $288,502. "You can say the value of the wood, but really that's too low," says State's Attorney Diggs. "Because what was taken was much more than just the wood. It's like taking an antiquity."

Old-growth forests also hold an economic power beyond the timber they produce and the environments they maintain. These trees attract people—millions of dollars in tourist money flood the Pacific Northwest each year. They are the main attraction.

The case was a lucky one, says Diggs in hindsight: the Shelton lumberyard reported the tree, the witnesses were forthcoming, the family property boundary obvious. All three conditions are atypical of timber-poaching cases, which are notoriously difficult to prosecute. "They don't leave fingerprints on these tree stumps," says Diggs, "and often very quickly after [the trees are] stolen they're broken down into blocks and sold." In Washington and British Columbia, it's not uncommon for wood to be posted for sale direct-to-buyer on websites such as Kijiji or Facebook Marketplace. Sometimes wood will be sold as firewood on social media. And if it's taken to a mill, its origin paperwork often proves to be expired or flat-out forged.

At the conclusion of the case, Reid Johnston (who maintained his innocence to the end, claiming that he'd been set up) told the *Seattle Times* that timber poaching would never cease: "There's plenty of wood in the national forest, and places they can steal."

Chapter 9

THE TREES OF MYSTERY

"I would go outside and it was all eyes on me."

—Danny Garcia

Approaching Orick from the north requires you to travel through some of the most imposing forested land in the world. Here you are guided from place to place through the belly of the woods, as opposed to skirting its boundaries. Danny Garcia had returned to Orick from Washington in 1994 along this route—his girlfriend, Diane, had given birth to their son, and the pair reconciled. He found work in a sawmill, where he learned to process the redwoods around him into shingles. Garcia had jobs in one mill or another, he had his familial roots, and he had a community—in short, many reasons to stay.

The first mill job he landed after returning to Orick was somewhat janitorial in nature: sweeping and shoveling sawdust and wood shavings after the processing was done. He worked a string of various sawmill jobs in the area over the following decade. His next job, for example, was pulling green chain, which entails gathering bundles of processed wood, sorting it by size, and moving it on to the next stage of the shipping process. Then he started driving a forklift and loading timber trucks. Loading was his favorite: he got to talk to plenty of different drivers, he was moving constantly, and the role was creative and required problem-solving: "You build your own load," Garcia explains. Plus, mill workers make connections and bond with those around them. In addition to the camaraderie, there was a lot of loyalty in those jobs; people stuck up for one another. When Garcia once accidentally crushed a coworker's hand, the man covered it up, fudging the nature of the injury to workplace health and safety officials. "He didn't have to do that," Garcia says now.

Around the same time, Sacramento transplants Lynne and Larry Netz bought a beige bungalow in Orick and settled down themselves. Lynne's son (Larry's stepson) Derek Hughes had entered 10th grade at the high school in McKinleyville, which he commuted to from Orick. Hughes had been diagnosed with attention deficit disorder in junior high and was taking Ritalin to help him stay focused, but in eighth grade he tried meth for the first time. It would eventually take the place of Ritalin, and Hughes would go on to use meth regularly for a decade.

Hughes was close to his mother and his older sister, Holly, but he had not had much contact with his biological father since Lynne moved the kids to California. Still, when Hughes was 16 he reconnected with his dad and went to live with him in Idaho. There he met a girl, and in 1999 the couple had a daughter. The child's mother was from West Virginia, so the pair moved east and got married. Eventually, however, the two split up; she moved back to Idaho with their daughter, and Hughes returned to Orick.

In the early 2000s, poaching from national parks across the entire United States was increasing. At the time, an investigation found there had been more than 800 thefts from archaeological sites alone each year from 1996 to 2003. Poaching—encompassing everything from deer and fish to timber and even Venus flytraps—was considered to be "very active" in at least 17 states. There was "nothing out there, besides air, that someone isn't taking," lamented a National Park Service representative.

In 2005, Chris Guffie (Danny Garcia's accomplice on the Olympic Peninsula a decade earlier) was caught poaching a log from a creek in Redwood National and State Parks, for which rangers found receipts totaling close to $15,000 in illicit sales. After they seized the wood and placed it in an evidence locker, someone broke in and stole "several hundred pounds" of the confiscated haul. In the end, a jury acquitted Guffie of grand larceny but found him guilty of vandalism. The park's chief ranger at the time had decided the park was being too light on thieves and poachers. In future, he decreed, they would come down hard.

By 2008, Danny Garcia's stable life had begun slipping. "There were several years there when things went downhill for me," he says today. "I went back to Orick, got caught in the riffraff of that community." Garcia and Diane broke up, and he moved into a small apartment above Orick's shuttered, run-down movie theater, across Highway 101 from the Palm Cafe and Motel.

Four years later Garcia wandered into that cafe, yelled at the owner, and pointed at a group of diners, saying he would kill them. He was charged with threatening with intent to terrorize, and given three years of probation under strict rules. After pleading guilty, Garcia was assigned a probation officer. The incident would play a key role in his being charged with poaching massive slices of burl off the tree near Redwood Creek.

Outside his apartment one spring midnight in 2013, Garcia met Larry Morrow, a man new to town who owned an SUV and was staying at the Green Valley Motel across the street from the

Palm Cafe. Garcia had offered Morrow $400 in exchange for using his name to sell wood to a burl shop. The pair drove until Garcia signaled for Morrow to pull over on the thickly forested shoulder. He asked Morrow to come back and meet him at the same spot a few hours later.

The land here was steep; Garcia's boots sank into the soft earth of the forest floor as he hiked up a small incline. Within moments he disappeared into the towering redwoods, some of them three meters wide, and began walking through the darkness, with only a battery-powered headlamp lighting his way. Now he was alone in the woods with his Stihl MS 660—a large chain saw with a 36-inch-long blade.

Garcia often walked "all over" the forest in his free time, he recalls: "I don't let bushes stop me, I'll make my own trail if need be. I see where they've poached burls, 50 years ago or longer." As he wandered, he scanned the ground for the soft, downy antler racks of Roosevelt elk, shed in expectation of new growth. Sometimes he even poached the elk itself. "If I need to get meat for my family," he says, "I'm going to get some meat."

The search for elk antlers is what usually led him to discover a burl. Garcia would notice trunks with large, protruding bulbs as he walked, or he might spy new saplings sprouting from a stump base. Occasionally, he remembers, he could almost sense a burl just under the surface of the ground, pushing up against the topsoil. "I've always been around wood, so I can look at the bark on a tree and tell you if it's good to cut, pretty much." Garcia likens himself to a favorite dog he once owned, who would catch an intriguing scent and take off after it. Hours later the dog would reappear at Garcia's apartment door, having followed his own tracks back home. Garcia remembered the location of good burls in somewhat similar fashion, and when he needed money he'd head out and follow his memory.

That spring night in 2013 he reached the tree he had in mind and fired up the chain saw. "The bark was probably 8 to 10 inches thick, and I could see the burl coming in through [it]," Garcia recalls. "It was probably one of the nicest burls I've ever cut in my life." After each slice reached the bottom, Garcia wielded the chain saw like a paintbrush, moving it steadily higher up the trunk. In the span of a couple of hours, he carved two vertical eight-foot lines into the trunk. Then, slicing widthwise between them, Garcia began peeling the burl out of the redwood tree.

He sheared off enough slabs to fill Morrow's SUV, then started to haul them out by hand. Doing so, rangers later speculated, would have been simply too ambitious: Garcia must have driven an ATV into the woods to transport that quantity of wood. But Garcia insists he removed it all by hand over multiple days.

On the forest floor surrounding the redwood, Danny Garcia left behind a thick carpet of sawdust.

The joint South Operations Center of the US National Park Service and California State Parks sits on the northern edge of Orick near the post office, and rangers at the SOC have become familiar with the intertwined lives of Orickites who steal wood. During the burl-theft crisis, a small group of people in town had earned the moniker "The Outlaws"; ultimately they would embrace that title themselves.

Terry Cook's property was known locally as "the Cook Compound," and SOC rangers suspected that it served as a kind of staging area for the group and its crimes. (Cook, Danny Garcia's uncle with the "big heart for people," called himself the "mayor of Orick," even though the town technically lacks one.) The most prominent outlaw was Chris Guffie: he was (and still is) known around Orick for his expressed goal to "rob the national park of their wood." Longtime local rancher Ron Barlow says Guffie was one of the smartest kids in town but that he chose a hard path, becoming known for "inspiring others to do wrong." Once, when reviewing footage from a trail camera that had been secreted in a tree, RNSP ranger Laura Denny thought she saw someone resembling Guffie with a chain saw in the park—but the person was sporting a woman's wig and sunglasses. "I'll tell you right now, I've taken wood out of the park, sure I have," Guffie told me over the phone in September 2020. "I've got no problem doing that. But I've never cut trees down; I've taken them up off the ground or whatnot."

After Preston Taylor had reported the poaching site off Redwood Creek, ranger Rosie White felt they might be close to catching and sentencing a member of The Outlaws. Having stashed security cameras in foliage near the tree, she waited to see if a poacher would return to the site and unwittingly get caught on video. Meanwhile a team of rangers began working their sources in and around town, interviewing some 20 burl-shop owners in Orick and along the coast.

During this time, ranger Denny traveled north to the town of Crescent City in the adjoining county of Del Norte, then south toward Eureka. Venturing into shop after shop, she interviewed the proprietors about the wood they bought. The process was straightforward: "When was the last time you bought wood, and from whom, and can we see the paperwork?" she asked. Some of the shops were inconsistent in what they could provide: they had a business license hanging in the window or by the register but kept no paperwork on their product. Others had no paperwork at all but could instantly rattle off their business-license number. Still others lacked documentation but swore up and down they had not bought park wood, claiming they could tell at a glance if a slab was legal.

Certain burl shops visited by Denny had everything arranged perfectly in their files, as if they'd been anticipating the moment when a ranger came knocking. Some bought and sold wood only from their storefront; others used websites such as eBay. All said they believed their stock had been harvested legally, through contracts with private logging companies or private landowners.

Because the subterranean burls harvested for veneers are massive and require uprooting an entire tree, burl shops typically traffic in the smaller ones, which appear on the trunk. Described by one ecologist as "a sort of wart," these burls come in two varieties: "gargoyle burls" sprout at the base of a tree and grow out and down, like gargoyles peering over the stone ledge of a chapel; the other type of burls appear as small globes farther up the trunk, bulging formations that grow out from the bark.

Most poachers know a burl-shop owner—their establishments are the middle ground between woodsmen and buyers. The upright owners require strict proof of provenance, much as an art dealer might for a painting or sculpture, but not all of them are honest. "I don't buy from local guys" is a common refrain in burl stores along Highway 101. During her interviews, Denny heard "I'm smart enough to know if the guy owns the wood" from one burl-shop owner—who later admitted he had no way of telling if the wood came from the park. "If I think it's hot wood, I won't buy it," another testified.

In practice, however, simple statistics still make burl shops the first stop when investigating a case of burl poaching: dozens of the stores thrive between Oregon and San Francisco, and there's only so much legally harvested product to go around.

A less common way of offloading poached wood, if a poacher has the skills or connections, is to forge the paperwork required by small mills. Some mills, knowing they can quickly sell old-growth lumber and hungry for the payoff, may turn a blind eye to missing or "lost" documents. By the time the sale occurs, it's essentially too late to catch the poacher: even if a mill is caught selling wood lacking paperwork, the wood cannot be matched to stump sites once it's been processed.

Instead, rangers tend to rely on anonymous tips incriminating poachers, hoping to intercept wood before it can be sold or carved. During Denny's interviews, an Orick burl-shop owner admitted that he had at times purchased wood from some of The Outlaws. In one instance, he said, a poacher he knew had stopped stealing and was working to "straighten himself up" doing yard work. The shop owner claimed that he could tell if a piece of wood had been poached from the park purely by its color and age. Steering Denny to some wood he claimed had been salvaged from water, not felled in the open, he said, "That's been rolling around in the river for a long time."

On May 15, 2013, Rosie White and Laura Denny parked at the head of the Redwood Creek trail and set out on the short walk to the cut site. As they stepped onto the trail, White noticed scuff marks on the ground, bookended by large piles of duff— the partly decayed leaves, branches, and bark that accumulate on the forest floor. Following the marks all the way to the site, they discovered a new cut across the front face of the previously damaged tree. The burl slabs that had been left a month earlier had been removed, and White wondered aloud if they might have been dragged out of the clearing by a vehicle like an ATV. The pair looked around: 200 feet of the hillside near the site had been disturbed, crushed by the weight of tumbling chunks of wood. The tree had been hacked up in five different spots, the cuts ranging from three to eight feet in height. Checking the hidden cameras proved fruitless: bright light from a headlamp had obscured the images, making it impossible to discern who had visited the site.

Four days later, Danny Garcia and Larry Morrow pulled into the receiving area behind a Klamath burl shop named Trees of Mystery. The pair unloaded eight burl slabs from the back of Morrow's SUV and showed them to the owner.

"Where are these from?" the owner asked. Garcia's response was vague, but he allowed that they had come from timber on family land near McKinleyville. After a bit of price haggling, the trio settled on $200 per burl. The owner took Morrow's driver's license to make a copy for his financial records, wrote out a check for $1,600, and gave the pair a pass to the shop's tour.

The following week, state park ranger Emily Christian received a text from an unknown number that said:

> hay u might want to see trees of mystery, they bought a Burl from Danny Garcia for 1600 cut into slabs a couple days ago

The next morning, Christian and Denny drove 26 miles north from the SOC to the Trees of Mystery burl shop. The enterprise, which is also a roadside tourist attraction, has earned a certain notoriety in northern California and with Pacific Northwest road-trippers: a five-story-tall Paul Bunyan on the site waves and talks to visitors, accompanied by a commensurately sized Babe the Blue Ox. In the years following the attacks of September 11, 2001, Trees of Mystery's status as a can't-be-missed attraction (physically, not metaphorically) prompted the Department of Homeland Security to identify it as a possible terrorist target.

On the day when Christian and Denny dropped by, the owner

admitted to having bought the burls for $1,600 from a man he knew only as Danny, who claimed to have harvested it on his grandparents' property. The owner showed the two rangers a copy of the check he had written, as well as Larry Morrow's license. On the showroom floor, he guided them toward four large slabs of burl that had come from the purchase, which he was now selling for $700 each; there were four more in the back. He promised the rangers that he would remove them from the display, and Denny took photos of them, including close-ups of the wood edges and grain pattern.

Denny and Christian hopped back in their truck and drove directly to the Redwood Creek cut site. Holding up the camera's review screen, Denny compared the slab photos with the tree cuts. The bark matched, as did the pattern of the wood grain. With these details confirmed, the pair had enough evidence to seize the wood from Trees of Mystery and begin tracking down its suppliers. Leaving Redwood Creek, the two rangers first stopped by Danny Garcia's apartment above the theater. Denny knocked on the door but no one answered, so they continued back along the highway to impound the burls.

Later that day, Rosie White returned to Garcia's apartment. Knocking on the door once more, she called out: "Danny, it's Rosie!" Again there was no answer, but this time she could hear a television on inside, as well as a dog barking.

A couple of hours later, White returned a final time, now buttressed by Denny and a Humboldt County sheriff. With Garcia still on probation after his threats at the Palm, a random check allowed them access to the apartment. Standing at Garcia's door, White knocked two times and identified herself while the sheriff stood in the alleyway outside. When still no one answered, White used a crowbar to pry the door out of its frame, then entered the apartment with Denny.

In a flash, Garcia's dog lunged at the two women. White tasered the dog, which ran into a bedroom, and she closed the door after it. There was no sign of Garcia in the apartment, though Diane—with whom he remained on good terms—was there taking a shower. "He might be at Terry's," a summary report quotes her as saying. "I know who steals wood in Orick. I know that Danny goes out at night with those guys."

But Garcia had in fact been on the scene: the attic in his apartment shared a wall with his niece's apartment, so he had simply kicked his way through the drywall, climbed into the rafters, and dropped down into her apartment, where he hid from his pursuers for the next three hours.

White commenced a stakeout the following day, parking her car behind the theater. When she needed to take a break, she asked the staff of the Palm Cafe, across the street, to keep an eye peeled for Garcia. The staff there stationed someone outside—but also reported to White that they had seen a man leaving Garcia's apartment building that morning, carrying a chain saw with a 36-inch blade.

Later that day, the rangers went looking for Garcia at Terry Cook's house. They parked their truck and walked down the long, unpaved driveway leading to the front door, passing stacks of equipment along the way that they suspected might be stolen. The yard of this Cook Compound overflows with the accumulated paraphernalia of a life spent cobbling together work: vehicles awaiting repair or being stripped for parts; stacks of wood to be processed; firewood for the stove.

The rangers knocked on the front door but were informed Garcia wasn't there. They searched six portable campers on the property and photographed the serial numbers of all logging equipment on the site; chain saws stolen from a Park Service storage shed the previous year were still missing. When they could find no serial number on an Alaskan mill, they seized it.

Back at the office, White found a voicemail awaiting her:

Hey, Rosie, this is Terry Cook. I just got home, and you went through my fucking yard and took a fucking Alaskan mill I had for thirty fuckin' years. You give it back or I'm gonna go up and I'm gonna start dumpin' trees, and you motherfuckers ain't gonna like it. I haven't been on your fuckin' park, and now I'm gonna start if you don't bring my fuckin' shit back. Fuckin' bitch!

Chapter 10

TURNING

"Never been illegal before, why should it be now?"
                                        —Derek Hughes

By the time he left school, Derek Hughes had become very good with his hands. He is lanky and so thin that his face appears gaunt, his features pronounced: large lips, a thin line of a nose, protruding ears. Using a chain saw came naturally to him, and he had taught himself the art of "turning" wood into bowls and other sculptures. Turning entails spinning chunks of rough-hewn wood on a lathe, which passes the wood dizzyingly fast along the sharp edge of a blade. Turners manipulate the wood until it smooths and softly warps, carving it into wooden bowls, mugs, and vases. The practice is as mesmerizing as raking sand in a Japanese rock garden, or watching water spill in a sheet over the edge of an infinity pool. Most turners sell their work to burl shops, which then sell the work to retail customers. Hughes estimates he could get about $30 for a large wooden salad bowl.

Lynne Netz's bungalow in Orick is a quick drive from where the mouth of Redwood Creek passes across Hidden Beach, opening into the Pacific Ocean. Growing up, Hughes often joined the rest of the town on the beach to gather firewood or go surf-fishing. He soon came to know wood as a staple of life in his town.

When high winds pass through the redwoods during a storm, branches, twigs, and dead-standing trees on forested slopes are pushed to the ground, picked up by water, and transferred downstream into the flux and flow of the sea. Floating along Redwood Creek, the wood eventually reaches Orick and often comes to rest on the bank of a river, where it is easily—and regularly—harvested.

The pattern of ownership along Redwood Creek, though, is inconsistent, a Morse code of public and private land. A redwood that was once rooted in park land might fall and be transported down Redwood Creek, washing up on private property where the owners can harvest it. But wood often travels all the way down Redwood Creek and floats out into the ocean, where ocean tides bring it back toward land and deposit it on Orick's Hidden Beach. Historically, plenty of townspeople heated their homes with that wood or sold it for fence posts, and the redwood trunks that drifted down to the beach would be scooped up, loaded into the backs of trucks, and either sold or stored in backyards for future use. "We all did that," recalls local rancher Ron Barlow. "It didn't belong to anybody. Well, it belonged to the state, I guess, but nobody cared about it."

Gathering wood from Hidden Beach was restricted in 2000, when the western boundary of Redwood National and State Parks was extended to the mouth of Redwood Creek and into the ocean. That year RNSP also introduced rules forbidding vehicles from driving on the beach or over the dunes. The plan addressed a problem that rangers near the park had been dealing with consistently: heavy traffic on some of the park's beaches, over which many people drove their vehicles to load up on wood or reach fishing vessels. Hidden Beach is home to the western snowy plover, which nests in dunes dotting the shoreline. Any disturbance to those dunes—being climbed by gasoline-powered vehicles, for instance—would exacerbate the bird's status as a recovering endangered species. Permits could be granted for wood harvesting and surf-fishing, but the new regulations made it difficult to harvest wood for anything more than a campfire. They also halted the granting of new fishing licenses and made it difficult for those with existing licenses to renew them.

The RNSP erected large iron gates to keep vehicles off the beach, but in so doing it also blocked access to a footpath popular with the community. The new restrictions (notably the requirement for permit applications) represented one more bureaucratic burden on a community that had already seen its access to wood monitored and taken away. Predictably, tensions flared in Orick once again.

Hughes watched the effects of that change unfurl around him. He remembers the anger in town as almost palpable. Residents accused park officials of "robbing them of their way of life," and of wanting to make Orick a ghost town (a charge still leveled today).

In the years that followed, fishermen found themselves ticketed for violating the beach laws. One local businessman filed a lawsuit against the park, then placed an ad publicizing his action in the *Eureka Times-Standard*: ORICK UNDER SIEGE, its headline declared. Another fisherman sued the park over his inability to make a living.

A "Save Orick" Committee was convened. In 2001 the committee hosted an "Orick Freedom Rally" meant to "highlight what has happened in our rural communities over the past 30 years because of the preservationist agenda of extreme green groups and land management agencies." Watching fallen wood clog the beach over time, local residents became incensed. The layer of washed-up wood became so thick in some places that it blocked groundwater drainage, flooding a nearby farmer's cattle pasture. Locals, including Terry Cook, began taking wood off the riverbank before it could reach the beach. Sometimes they waited to retrieve a log until it had floated beyond the park's property line; sometimes they did not.

Then things got worse.

In 2003 the National Park Service revoked overnight camping and wood gathering at Freshwater Spit, a sandy beach just south of Orick that was often stacked three-deep with RVs and trailers. (In Park Service eyes, the thousands of trailers jamming scenic Highway 101 had created an "aluminum blight.") The Freshwater Spit decision became a local flash point: to this day many Orickites invoke the meeting at which Lucille "Mother of the Redwoods" Vinyard, the outspoken leader of the Save the Redwoods League, allegedly pressured the Park Service to get rid of the campground altogether. Yet Barlow, for one, agrees that camping had become "out of control" on the Spit, where the overpacked parking had made it impossible to visit the beach.

The camping decision enraged townspeople in Orick, who feared their businesses would be affected simply because "parks people didn't want to see trailers." James Simmons, the owner of Wagon Wheel Burl, reliably sold one burl table a week to camping tourists, he reported, and the items had become his bread and butter.

In retrospect it's not difficult to see their point—especially today, when preservation purists appear to pander to visitors who live outside the region and who want aesthetic beauty along with their wilderness. (The US Forest Service rents luxury cabins on its land, while Parks Canada offers "glamping" in heated yurts.)

When Orick residents attended a public forum hosted by the park, "we about blew the roof off that place," remembers former logger Steve Frick. "We lived on [that tourism], we needed it. Hell, half the stores in town, that's where all their summer money comes from. They couldn't make it off loggers [who] couldn't afford to go out every night."

As tensions in the town escalated, protests and pickets popped up along the highway. Then death threats were made against rangers, and a pipe bomb was discovered in an outhouse. The Park Service called in a SWAT team.

In a letter to the Department of the Interior, community members requested that RNSP reinstate some camping at Freshwater Spit and lift its latest restrictions on wood gathering. They asked Interior to appoint a federal mediator to reconcile the town and the park. "The final blow to the community of Orick is just around the corner," warned an ad in the *Humboldt Times*, detailing the proposed changes and imploring readers to write their government representatives. The newspaper ad concluded with an appeal to the conspiracy-minded: "P.S. As you read this, a federal SWAT team sits, ready, at RNP South Operations Center. The siege is real!!"

Chapter 11
BAD JOBS

"It's all over the world, it's where you're at too.
Don't think for a second it's not."

—Derek Hughes

L iving in Orick, Derek Hughes entered a pattern of gig work:
he worked the night shift at Greyhound, just as his mother
had done. He mowed lawns outside businesses for the town's
Chamber of Commerce. He helped renovate the house across
the street from his own.

This work arrangement was not unique to Hughes, nor was
it a temporary employment situation. Orick's poverty rate was
26 percent in 2021, and the town's destitution was difficult to
overlook. Many of its houses had not been maintained consistently, standing as sharp indicators of stalled time. With only a
gas station and a small market selling food—their prices easily
doubling those found in large grocery stores in cities nearby—
Orick is a food desert.

How can this town on the edge of one of America's most
iconic natural wonders not be flooded with tourist money? Why

are its streets dead even in high summer? On the surface, the answers boil down to small, simple details: Orick's Chamber of Commerce has struggled to mount street-side refurbishment programs, such as installing flower planters and signage; home and business owners have not invested in renovations; there are few vacation rentals on offer. Social planning also fell through the cracks.

Civic decay and economic hardship form a sort of feedback loop: one study notes that "economic and social changes in the last 50 years have led to communities where high rates of poverty are concentrated in neighborhoods with crumbling infrastructure." This is apparent in Orick: in 2014, a small Mexican restaurant closed down, and in 2019 the Palm Cafe and Motel followed suit. That left the Snack Shack, a takeout hut with picnic tables that serves burgers and fries with logging-themed names (the Log Deck, the Driftwood Tots, the Flaming Redwood), as the only restaurant in town. The theater has closed indefinitely, and the small school, with fewer than 100 students, has narrowly escaped being shuttered numerous times over the years.

After his mother died, Jim Hagood took over operations at Hagood's Hardware, his parents' business in the heart of town. Before that he did whatever he could to earn money for his family: he served in the military, then headed out to the woods, logging or driving logs to markets in Eureka or Del Norte County. He and his wife, Judy, raised a son and a daughter, and they were (and still are) active members of the town's volunteer fire department.

Jim is now one of Orick's longest-tenured residents. He lives in the house behind Hagood's Hardware, keeping a separate shed fully stocked with nonperishable food items to survive imagined future disasters. He meets regularly with Joe Hufford, who runs a close second in town tenure and whose grandparents started the first-ever sawmill at Redwood Creek; Joe is a third-generation logger.

The Hagoods spend their days in the shop, conducting their lives on the other side of the plate-glass windows fronting Highway 101. The store is a menagerie of memorabilia: newspaper clippings, photos, antique equipment, and Judy's quilting projects, which she donates to a children's hospital. They serve pie to visitors on small paper plates with white plastic forks, handing them over stacked back issues of *AARP The Magazine*. A thick layer of dust covers most items on the shelves, and a faded Boy Scouts of America flag hangs beneath moldy ceiling tiles threatening to fall. There's firewood for sale near the entrance; on the floor nearby stands a plastic bin overflowing with Jim's homegrown lemon cucumbers. Wrapping a handful of the fresh vegetables in used grocery bags, Jim encourages me to eat them on the road.

Judy perches on a stool behind the till, invisible until you hear her, holding a walkie-talkie that spurts to life with static, voices crackling back and forth across the county. Jim usually holds court at a large folding table, the surrounding walls covered in photos celebrating Orick's boom days. Above the original front door, now impossible to open, spiked logging boots and hard hats hang from the ceiling. By the time I leave, my boots and glasses are covered in a layer of sticky dust.

It's easy to get ensnared by the past in a memory palace like Hagood's. The town's halcyon days—when Orick boasted five mills and 22 dairies, and more than 40 sawmills dotted the county—are remembered fondly. "We never thought of it as a logging community," Hagood says now. "It was just a close-knit town." Joe Hufford's mother, Thelma (not to be confused with Terry Cook's mother), wrote a weekly dispatch for the *Arcata*

Union, featuring stories about the annual Banana Slug Derby in the park and poems extolling Orick's natural beauty. Today the park's management is seen as wasteful, inefficient, wrong. "We have about as much use for government in this town as we do a dose of clap," says Jim Hagood.

Orick's contemporary social troubles are inextricably linked to the waves of unemployment that washed over the region 30 years prior. Layoffs, author and social worker Joseph F. Madonia writes, deeply affect the self-esteem of those who lose their jobs; in fact, they are a source of trauma. "Whole lives are tied with jobs," Madonia's research concluded, and for those who had lived in company towns or held positions with a single company their entire lives, losing their employment was "a profound shock. The job and the company are so closely related to their identity that the first reaction is one of crisis."

That crisis is still playing out in Orick and other communities across the Pacific Northwest. In studying the effects of unemployment on California timber communities in the 1980s, researcher Jennifer Sherman found that low self-esteem often led to increased substance misuse, domestic abuse, and crime. Logging, writes Sherman, was "the axis along which the pattern of lives were organized...there is a significant meaning to work, and without it many difficulties ensue."

Much of this meaning is rooted in the occupation's tasks: loggers work no matter the weather, in extreme heat, and during long rains. Because of the terrain, the weather, the environment, and the immense volume of work, the danger inherent in logging had, at least in part, forged the identities of those who undertook it. Part of the pride that came along with the work arose from the knowledge that previous generations had lost their lives to the forest. Thus the central meaning of their lives, as sociologist Clayton Dumont once noted, "stems from having literally invested their own blood in timbering occupations."

Though opportunities for work existed elsewhere, a core group of Pacific Northwest residents felt so connected to the region that they refused to move after the industry declined. They fought for their jobs in the ways they knew how, but that struggle would soon sour into feelings of pessimism, then anger. As a result, studies have identified timber poaching in the Pacific Northwest as part of a "cultural practice" that reinforces a once-shared heritage, providing a pathway into community acceptance.

Researcher Sherman's work later concluded that shame, guilt, illness, stress, and addiction were all symptoms of the unemployment that descended following deindustrialization in the Pacific Northwest. "In rural areas," she writes, "industrial decline can be particularly devastating because it results in the loss of a 'way of life,' which is centered around the work in the declining industry." In other studies, many participants said they didn't want to rely on emotional or financial support for too long, because it reminded them of their unproductiveness. Sherman watched unemployment cultivate turmoil: the structure of families often changed, for instance, if one parent moved for work, or if parents divorced in the aftermath of a job loss. The effects on family ties were undeniable: "The father works in the woods," explained a teacher in Humboldt County in the late 1990s, "then his boy has the dream." That tradition was lost when work dried up.

Many loggers identified with a job that no longer existed— one that had promised prosperity as they were growing up. It became difficult for them to find their place in this new era. Many of the unemployed reported feeling emasculated as their wives went to work. "General disharmony" occurred in families, Sherman writes, leading to power struggles, pervasive anger, and self-destructive behaviors such as substance misuse. Eighty percent of the jobless surveyed said they were more agitated and more frustrated than normal, and that work was the only thing capable of banishing those feelings Eventually the psychological strain of unemployment became so extreme that some respondents felt poor mental health weighed on them more than financial concerns.

This led, eventually, to what the Bureau of Labor Statistics calls a force of "discouraged workers"—those who want to work but can't find a job and finally just stop looking. Most often these workers are categorized as "unskilled" or "undereducated," and they tend to live in high-unemployment regions. This would not be assuaged over time: many of the jobs lost in the late 20th century were not replaced, or at least not with jobs that guaranteed stable, full-time work and a fair income. Instead, a plethora of "bad jobs" arose: temporary or contract work with low hourly wages, or nonunionized service jobs with no promise of minimum hours or consistent pay. Bad jobs are held primarily by those lacking a bachelor's degree: "They do not get to live in the fast-growing, high-tech, and flourishing cities, and are assigned jobs threatened by globalization and robots," explain Princeton economists Anne Case and Angus Deaton, who have researched the topic extensively.

Men without four-year degrees lost 13 percent of their purchasing power from 1979 to 2017. The concomitant loss of pride and sense of belonging to a workplace has made for "lives that have come apart and have lost their structure and significance," Case and Deaton also note.

Automation, globalization, and increased education requirements—compounded by failures in government and institutions—have given rise to a generation of disconnected and fearful people. The number of men who have dropped out of the labor force and stopped looking for work has quintupled since the 1950s. The result is a form of community trauma deeply felt in many rural areas: intergenerational poverty, long-term unemployment, degraded environments, disconnected social relationships, and destructive social norms.

Present-day Orick and Forks took shape in the shadow of this trauma. Both towns occupy prized tourist real estate in the country. But, "there's no occupation here," says Jim Hagood. "There's no work. The logging is gone." Rather than seeing its tourist trade boom, says the president of the town's business chamber, Orick has become "disinvested": people left and took their savings with them instead of starting tourist-oriented businesses.

———

The emotional upheaval of unemployment also took place alongside a spread in methamphetamine use in rural areas. While much of the country was distracted by the crack epidemic in the 1980s, meth use trickled into rural counties of northern California and the Pacific Northwest. Drug misuse experts working at clinics in the Haight-Ashbury neighborhood of San Francisco had sounded alarm bells at the time, saying that meth was following motorcycle gang routes in the state, and that the Hells Angels and Gypsy Jokers were dealing in remote rural areas.

Meth would become ubiquitous across the Pacific Northwest. By the early 2000s it was considered northern California's most significant drug threat, with one in four US meth users living in the state, where the drug was being manufactured in homes and backyard labs. Reports from California's justice department noted its prevalence in domestic-abuse cases; its ease of production, its ready availability. In 2004, Portland police responded to hundreds of complaints about meth houses in the city. In Vancouver, low-cost hotels in the city's notorious Downtown

Eastside saw their clientele slowly change from loggers on break to drug users and folks with mental illness.

Drug treatment at that time did not often focus on the link between meth and work, but meth had originally been developed as a worker's drug. During the Second World War, amphetamines were a relied-upon stimulant to keep troops active. The Japanese called it *senryoku zokyo zai,* meaning "drug to inspire the fighting spirits." Over time, meth use gained favor among long-haul truck drivers and mill workers, whose extended shifts required them to remain quick and alert.

"We did see a lot of people who were using [meth] as a way of coping with harsh employment," says Mike Goldsby, a sub-stance-abuse counselor from Humboldt County. Meth also had the advantage of being cheap to produce. "When those [logging] industries started to decline," he says, "meth was inexpensive compared to other drugs, and it was often a way that people would kind of drift into [drug use]."

As government at all levels worked to combat the spread of methamphetamine in the early 2000s, some states began to require prescriptions for formerly over-the-counter cold medica-tions containing pseudoephedrine, a key ingredient of meth. By the end of that decade, however, the opioid epidemic had started to dominate coverage of the drug trade. Meth might have left the limelight, but it hardly vanished from the scene. It has since morphed into a big business—no longer cooked in backyards as often, but instead trucked to dealers through cartels.

The relationship between drug use and unemployment is com-plex. Does someone take drugs because they are unemployed, or are they unemployed because they take drugs? One of Jennifer Sherman's studies described a cycle of poverty, desperation, and self-hatred leading to substance abuse. There is a pernicious feedback loop: unemployment rates are high among meth users, drug use contributes to prolonged unemployment, and users are likelier to relapse if they don't have a job.

Jim Hagood likes to quip that there are three churches in Orick: "Catholic, Baptist, and crystal meth." Other voices throughout Humboldt County echo his observation about broader drug use in town. A man who works in addiction services in Eureka says Orick is sometimes called "Oxyville" because so much OxyContin is being dealt on the streets. The town's Green Valley Motel is distinguished by its peeling blue siding and long-term tenants, its parking lot brimming with discarded furniture and abandoned vehicles. A man was once found slumped over a railing outside, a fentanyl patch drooping from his mouth.

Meth use has taken a harsh toll on small communities across the Pacific Northwest in the past 20 years. A 2019 report found the region "drowning in methamphetamine" and linked its use to the opioid crisis: meth is a surefire antidote to feeling drowsy, so it is marketed to users of heroin and other opiates. County officials in Washington say that needles found on streets are often assumed to be relics of heroin use, but in reality they are most often used to inject meth.

In November 2020, the National Institute on Drug Abuse reported an alarming increase in overdoses involving a combination of opioids (such as heroin or fentanyl) and stimulants (such as meth). A study published in the *International Journal of Drug Policy* likewise signaled that unpredictable product quality had prompted many opioid users to switch their drug of choice to methamphetamine. Thousands of Americans now die from methamphetamine use each year; the number of people overdosing has tripled in the course of a single decade. "What I know is that meth is everywhere," says Derek Hughes. "Everywhere. Even in places that you would think it's not. And it's thriving."

In the Washington counties covered by Olympic National Park, methamphetamine deaths increased an astounding 442 percent from 2004 to 2018. Of the 531 people who died in Washington in 2018 due to methamphetamine, 77 percent were White and over the age of 47—a demographic that includes many people who grew up in the state during the Timber Wars, watching local communities get squeezed out of the logging industry. In Humboldt, meth accounts for a quarter of all drug overdoses. Over time, its use has risen in lockstep with homelessness in the Pacific Northwest.

In her 20-year career, Forest Service investigator Anne Minden became certain that timber poaching was intimately tied to drug use. "A lot of these individuals, unfortunately, are drug-addicted," she says. "I've seen drug addiction in, I would say, 90 percent of the cases I've worked involving cedar or maple theft." Her observations are similar to those of RNSP chief ranger Stephen Troy, who says that Orick's poverty, paired with "serious addiction here with meth," is a prime motive for poaching. Both sentiments were echoed by dozens of people I interviewed, who place the blame for poaching on "tweakers," "meth heads," and "druggies."

But, "a hurt is at the center of all addictive behaviors," writes addiction researcher Dr. Gabor Maté, whose work has strongly shaped recent trends in drug treatment. Applied to the Pacific Northwest, his assertion gives us a keen insight into the rampant use of hard drugs: they are used not simply because they make it easier to work, but because they are palliative, effective at softening a pain. Unquestionably there are consequences to using meth—but what if those consequences pale in comparison with how life feels without it?

Mike Goldsby's counseling now focuses on childhood trauma. "We understood that most people would look at an addict of any kind (but especially a meth user) and say, 'What the hell is wrong with you?'" Goldsby explains. "But we would look at [that same addict] and say, 'What *happened* to you?' What's become apparent is the desire for the relief that drugs bring."

Still, being associated with drugs is as stigmatizing in a small town like Orick as it is anywhere else. Substance abuse is spoken of shamefully, and users are judged as "not contributing" or "worthless." When Danny Garcia and I first started communicating in early 2020, for example, he denied using meth. But paperwork tied to his court case indicates "regular use of controlled substances" and categorizes his drug use as "severe."

Derek Hughes deems it unfair to say meth users poach wood in order to get drugs: "We have bills to pay. We're just like anyone else except we live in the middle of nowhere, there's no jobs, and they don't want to hire us to prevent wood poaching." People assume poachers do drugs because "we're up at night working," he says. "Well, we're up all day, too."

Hughes was one of the few principals in this story willing to be frank with me about drug use. From our early exchanges on, he voiced his desire to quit even though [meth] helps him focus. "It's a bad thing and unfortunately I have the habit," he conceded. "But I couldn't be without it, because I don't want to take Ritalin anymore." Our discussions often turned philosophical: "There is a disconnect between people who do it and people who don't. The disconnect is one set of people think they're better than the other. But I have morals. I have scruples. I'm a damn nice person once you get to know me."

Chris Guffie, now dubbed "the Redwood Bandit," bristles at the notion that people poach timber for meth. "You can go to a doctor and become a drug addict," he points out.

"But let me ask you this," Guffie adds later. "When you go out and work, you get paid, right? You take care of yourself? The basic person when he gets his money, he takes care of his needs.

And what he does after his needs are taken care of or whatnot, it's his business."

———————

Notably, one of the clearest links between burl poaching and meth use comes through an episode of A&E's hit reality show *Intervention*. In one episode we follow Coley Town around Ferndale, a city in southern Humboldt County. The episode portrays Town as a sensitive, dedicated husband and father who happens to be addicted to meth. His early life had been difficult, stretched to the breaking point after his parents' marriage crumbled, and he was deeply affected by his mother's own addiction. Town worked as a logger for more than a decade. "It was serene work," he tells the camera. "Other than my chain saw screaming and the trees squealing."

When we meet Town he is unemployed, but convinced that burl poaching will land him enough money to escape poverty. The audience watches as he smokes meth in his garage, then heads out burl hunting in broad daylight. "This is burl country, dude, I'm telling you," he reassures a friend as the pair drive a small truck toward the forest.

We follow along as Town hikes into the woods (the precise location undisclosed) and are just steps behind when he shouts down to his friend that he's found a prime burl clinging to the stump of a logged redwood. Viewers witness Town scaling redwoods with his chain saw, slicing off burls midway up their trunks, then throwing the forbidden wood downhill to be loaded into his truck bed.

Town explains (incorrectly) to us that burls are a cancer on the tree, but that they're marketable and net big money. Meanwhile the meth makes him feel invincible—"like I could move a mountain with my finger," he says. The camera follows Town through the woods for nine hours. "You gots to get up here, dude," he calls out at one point. "I found the biggest one I ever seen…am I dreaming? We'll never work again! That's beautiful…it'll get us 20, 25, 30 grand!"

"We have not been able to get him off the burl," his wife tells the camera. "Meth is hand in hand with this crazy obsession he has right now." Town echoes her thoughts: "Not only am I addicted to meth, I'm addicted to burl, curl, swirl." He sees the burl as both a symptom of the addiction and an opportunity to save himself. But over the course of three months, Town doesn't sell any burl at all.

One day, the bed of his truck laden heavy with chunks of wood, he drives the haul to Burl Country, a shop outside Eureka that I have passed by numerous times. Its large sign beckons from the highway. After assessing the poached wood, the shop owner offers Town a paltry $500. Town is being undersold, he claims, because he has "the tweaker look." In the end he drives off and brings the burl home.

Reviewing the episode, the *New York Times* described Coley Town as "a deranged American frontiersman." But more nuanced reviews appeared online: "I come from a poor Pacific Northwest logging town, raised and surrounded by loggers and mill workers who snorted crank like other people drink coffee," one fan wrote on intervention-directory.com. "Coley, his lifestyle, the way he uses, is intimately familiar to me in a way that actually makes me uncomfortable to watch."

Town looks familiar to many who live or grew up in the Pacific Northwest. One former meth user I spoke to remembers sitting in a dealer's Washington apartment one night when a man walked in holding a freshly cut slab of maple. He had plans to turn it into a guitar.

———————

In Humboldt especially, drugs and the drug trade hold a unique place in the wider culture: they are at once the savior and captor of the county. Part of northern California's "Emerald Triangle," Humboldt's economy has been fueled by marijuana sales for decades, both on the black market and on the (now) legal market. The roads into the county are lined with billboards advertising marijuana companies, and old tourist attractions have been converted into dispensaries. The economy is so driven by growing cannabis that public-radio commercials advertise the large barrels used for picking during harvest: "On sale now at Costco!" (Donations for the station are solicited this way: "Where else are you going to spend that big payday once the 'golden bud' is cashed in?")

Local lore mandates that you never ask a Humboldter what they do for a living. "It's a special brand of etiquette," journalist Lisa Morehouse wrote for *The California Report*. To avoid crop-picking fingers made sticky with marijuana resin, she explains, "During harvest you don't shake hands; you hug with your arms wide and palms pointed up."

Big business has now learned what so many backyard growers have known for decades: the environment that grows imposing redwoods is humid, damp, shaded, and perfect for growing weed. Some of the largest swaths of illegal deforestation on public lands in Humboldt County have been perpetrated in the service of marijuana; researchers and park rangers have confirmed thousands of criminally logged locations in northern California that were converted into large-scale marijuana operations. They compare the damage done to the devastation caused by poaching elephants.

Yet the booming weed economy has not stepped in as an economic savior for smaller towns like Orick. In practice, it has left the region struggling with yet another shift in its economy:

the transition from clandestine to legalized distribution has not been as simple as moving pot production into the daylight. The Netflix series *Murder Mountain,* for instance, chronicles the outlaw groups that run large-scale illegal "grow-ops" in an area not far from the back-to-the-land communities near Garberville. It shows streetlight poles and community message boards in Eureka and Arcata, papered with posters for missing persons (northern California leads the state in cases of missing Indigenous women.) If you want to get away with breaking the law, few places can compete with Humboldt County. Its violence has morphed, turning in on itself as a failed utopia. The county attracts the underground, the seedy, the environmentally obsessed, and the transient. "The peace and love went out with the '70s," says one resident in *Murder Mountain.* "Now it's about the dollar. Here is where you can find the last vestige of the Wild West."

At the same time, it has become increasingly expensive to live in Humboldt. Many of the county's long-term inhabitants have been priced out of their communities. Crumbling housing stock in towns like Orick costs far more than would-be denizens can afford. The town's Chamber of Commerce is concerned that homes will fall to ruin before anyone shells out the money being asked by the overheated market. Thus Orick finds itself ensnared in a vicious circle: its reputation for drugs and unsightly property deters anyone who might want to invest in making it a permanent home or a place where tourists might want to stay.

Jennifer Sherman concludes her pathbreaking examination of the Golden State's lapsed logging trade with a nod to politics: from the perspective of many interviewees, she writes, the government "prioritizes the interests of urban liberals, such as the reviled environmentalists who are blamed for current impoverishment. Most residents feel that neither side has much interest in their material well-being and economic sustenance."

But with the right wing focusing on moral and personal issues, notes Sherman, it succeeds in hitting closer to home: "When your hunting rifle is a major source of your ability to engage with your children, reproduce your unique culture, and provide for your family, gun control will likely seem a serious threat."

The same could be said for an ax.

Chapter 12

CATCHING AN OUTLAW

"That tree was under surveillance."

—Danny Garcia

On May 25, 2013, only a day after hearing Terry Cook's angry voice bellowing into her voicemail, Rosie White arrived at work to a phone message from Danny Garcia. He'd like to talk, he said, and agreed to visit her office. Just after lunch that day, he sat down and spoke with White for over an hour.

Sitting across from her, as noted in park documents, Garcia admitted to being familiar with the eight burl slabs confiscated from Trees of Mystery. He had first seen the wood, he maintained, at the gateway to a local gravel lot about a mile from the Redwood Creek trail. He described the wood's distinctive bird's-eye swirl from memory and, according to a later report from White, implied that he had trimmed the slabs himself. White then led him to the SOC storage cell where the Park Service was holding the seized wood, and Garcia confirmed that these were indeed the pieces he'd seen at the gravel lot. Garcia's mere familiarity with the wood, however, hardly justified an arrest. The rangers needed tangible evidence that would place the contraband, and a chain saw, in Garcia's hands.

Two days later, the haul from Redwood Creek was joined by even more wood, which park rangers had impounded from a shop in Orick named Burl Bill's. The wood had been stacked on the store's porch, and park rangers were able to match its grain with that in the photos taken at Trees of Mystery. The shop's owner said the burls had appeared "in the middle of the night" and that there was no paperwork for them. "I asked around about them," he said. "And the rumor is that Danny Garcia dropped them off."

Yet still the rangers lacked enough proof to secure a conviction. So they continued to investigate new poaching sites, hoping they might be able to charge multiple members of The Outlaws with poaching.

The rangers had begun to suspect that poachers were eavesdropping on their radio communications, enabling them to track the rangers' movements and avoid the park when they were around. Park Service officials likewise chafed at their failure to remotely photograph poachers in the act: the images captured by cameras secreted in trees were easily distorted by bright headlamps or flashlights, and the cameras themselves were routinely stolen from their hiding sites. So a Park Service team set out to adopt more advanced detection methods. Working with researchers from Florida International University and California State University, the team embraced LiDAR (light detection and ranging), a technology that allowed them to scan the forest from above. LiDAR let the rangers pinpoint the location of the park's most poachable trees. The team then strategically placed cameras and certain other monitoring devices nearby.

The rangers also turned to another new method: magnetic sensor plates, hidden on the forest floor, that would spring to life upon detecting the dense metal of a chain saw. Two of these plates, costing around $10,000 each, were buried at previous poaching sites that rangers believed would be revisited. Whenever a sensor plate was activated, it would surreptitiously alert the SOC. Since being installed, however, both sensors have remained silent.

In late May, rangers Rosie White and Laura Denny requested that the National Park Service's special investigator, Steve Yu, come in to help. Yu was based in Yosemite, but he traveled north and spent the summer at Redwood to work the case. Timber poaching, he says, "had brushed up against my consciousness, but I wasn't super up on it." When he arrived, he noticed that Orick was not a typical gateway community. "Well, most gateway communities, in my experience, they cater to the visitors." But Orick, "it's kind of a sad little place…meth is everywhere in that town."

Yu and the rangers sat down in a conference room filled with whiteboards and began mapping out the town's social dynamics. "We draw the connections, see what holds, see where the missing information is," he says. "At first, it's like drinking water from a fire hose, right? But once you have that information on the board, then you see what you need to focus on."

It was apparent to Yu that Chris Guffie was a central Outlaw. But the Park Service goal was to send a message by actually convicting a poacher, and they had a stronger case against Garcia: "We sidebarred Guffie," Yu says. "Guffie, as I recall, is pretty darn smart. That's his identity. He's not going to stop." Instead, they decided to focus all their resources on proving Garcia's guilt.

Yu started accompanying rangers on their investigation and interviews around town. "It was this real cat-and-mouse game between Chris Guffie and the rangers, Garcia and the rangers. It was this ongoing thing," he says. Near the end of June, the team began to suspect that some of the wood in Orick's burl shops and yards originated from a stretch of Redwood Creek that lines the boundary of Redwood National Park, as opposed to sites deep in the woods. When White, Denny, and Yu visited the creek one day, they found two redwood logs in the water, with metal cables anchoring them to the shore. Some sections of the logs had already been cut off. Scattered around the riverbank were gas canisters, sawdust, and a tailgate that had fallen off a truck.

The rangers spent the summer of 2013 investigating a spate of illegal burl sales throughout town, including one instance where a slice of burl was found hidden behind the Green Valley Motel, where Larry Morrow was staying. Another slice turned up at a storage facility, whose owner told investigators he had overheard someone talking about the hidden burl.

In the meantime, Garcia attended regular meetings with his probation officer, a requirement linked to his 2012 outburst at the cafe. Throughout that spring and summer, says Garcia, he was visited so "constantly" by park rangers that it started to feel like harassment: "Seemed like every few days they were there— searching, looking, searching, looking." They never found any wood in his apartment, but they did seize some driftwood from his truck, saying he lacked a permit to gather it from Hidden Beach.

While the Park Service's fruitful investigation into Garcia's case wrapped up in the summer of 2013, securing a warrant for his arrest would take months longer. Rangers had to wait until the spring of 2014 for their case to be processed through Eureka's court system. Then, rather than busting down the door of Garcia's apartment, they simply attended one of his probation meetings in April and charged him with grand theft. In May, Garcia was convicted of felony grand theft, vandalism, and receiving stolen property. He was fined $11,178.57 for poaching redwood and selling it to burl shops. Larry Morrow, who had accepted a plea deal in the case, was given only three years' probation.

In the lead-up to the case, ranger Laura Denny had arranged for a forestry expert in Arcata to visit the burl-poaching site. The expert, Mark Andre, is a forester for a local consulting firm, as well as an expert in assessing and valuing timber. In addition to calculating timber volumes, he measures tree height, diameter, and quality. Andre noted that chain-saw cuts had stitched their way around the redwood's 10-foot-diameter trunk, exposing its heartwood. Although the tree remained standing, it was at risk of disease and rot—indeed, its lower section was rotting already. After measuring the pieces of burl held in the SOC evidence locker, Andre estimated the damaged tree's total value as close to $35,000.

When it came time to determine Garcia's sentence, the judge remarked that "Mr. Garcia's criminal behavior is related to what I believe is drug addiction issues...looking at your entire history, it does appear that drugs are the root of your problems." Then the judge emphasized the deeply serious nature of the crime: "I think it is fair to say that living on the North Coast with the beauty of the redwoods all around us, we don't necessarily appreciate it in its entirety. I think you, Mr. Garcia, were opportunistic when you damaged the tree...the trees that are preserved for the citizens of our state and nation and the world, in fact."

The judge also noted the crime's unique nature: "There aren't any statutes in our state laws...that really apply specifically to this case. I do think that your sentence, Mr. Garcia, will send a message to the community that hopefully others will not engage in the same behavior.

"Mr. Garcia, what I do want to say to you finally is what I see.... I see a person who has committed offenses because of what looks like chronic drug use. You can't change any of that, but I think living at a different location will assist you in making meaningful change, and I hope that occurs for you."

Garcia, for his part, rejects the characterization. "I cut that burl because I needed rent money," he reflects from his home in Eureka today. "I've had my issues over the years, but [meth] wasn't leading me to cut that burl."

Appealing for a lower fine, Garcia detailed how press coverage of the case had made it hard for him to get a job. "They made me out to be the worst, and it bothers me. There's a lot of [burls] out there, and I didn't kill the tree," he insists. "That tree ain't dead. As for the damage it does to them, I do and I don't think about it. It's always in the back of your mind. I don't feel it's right what I did, but then again it's not hurting the tree as much as they say it is."

Standing in his yard, Terry Cook says Garcia took matters too far when he cut the redwood so extensively. "I said, *Do that again, you fucker, and I'll take you down and turn you in myself,*" says Cook. Cook's partner, Cherish Guffie (Chris's ex-wife), nods in agreement: "We were mad at him for that. Everybody was, because that was stupid and dumb, okay?"

Garcia was jailed in mid-May of 2014, then released after accepting a plea deal. He was ordered to pay the full amount of the fine in installments, and to stay away from the national park. On October 22, 2015, the Park Service used a wood chipper to destroy the redwood burls that Garcia had poached.

Chapter 13

IN THE BLOCKS

"Show me your friends, and I'll show you your
future."
                                        —Derek Hughes

The wood turner Derek Hughes continued living at home
into his late 30s. There wasn't much of a career ladder available to climb, and he found it difficult to grasp hold of anything
permanent or stable. Hughes's home life could be rocky, according to his mother, Lynne Netz; his stepfather, Larry Netz,
experienced mental-health challenges that only escalated as
Hughes grew older, and Hughes was unable to save up enough
money for the first month's rent and a security deposit on his
own place in another city or town.

The Netz residence is heated by a woodstove, and Hughes sometimes traveled to a friend's nearby property to cut oak and madrone
to fuel it. Lynne Netz prefers the heat off those species because
they burn hot and slow; she can put a log in the stove before she
goes to bed and find coals still burning in the morning. But Hidden
Beach, only a couple of miles behind their house, is easier to
access—even if the old, dry redwood waiting there burns quickly.

Hughes's family house is in a community everyone calls "the Blocks," nestled just behind the businesses that line Highway 101. Lynne calls the neighborhood "the projects." Many homes in the Blocks appear haphazard: whereas some are being re-habbed, others have yards filled with stacks of chopped wood, or parked cars, or tarps covering old machinery.

Hughes first started poaching wood from Hidden Beach with Danny Garcia around 2010. The two had met through a net-work of acquaintances that circled around the Cook Compound. Though Garcia is about a decade older than Hughes, they became fast friends in adulthood. "You know, two adults got to make some money," Hughes allows. "[So] that's what we did."

Garcia began taking Hughes along on his late-night wood-poaching forays, then Hughes started going out on his own. Working in his street clothes, he parked his vehicle as close to the beach as possible and hauled the wood by hand into the truck bed. He headed out rain or shine. With a flashlight and a chain saw, he would cut into the logs that dot the sand, cleaving large chunks off their sides. Sometimes the remains left behind resembled benches—straight-backed seats on which to watch the sea.

At first, poached burls fueled both the Netz woodstove and Hughes's wood-turning practice: "It's pretty easy to get pieces of wood that you can make a bowl out of," he says. "I figured I could make some money at it if I got good." Ultimately, though, he turned his eye to larger hauls he could sell to local shops and artisans for bigger and more immediate payoffs. That wood made it into the system sometimes as art, sometimes as shingles. "It's not hurting a damn thing," says Hughes of taking the wood. "And you know, people would stay out of the woods if they gave us the beach!

"Nobody goes out to do this saying, *Oh, I am going to go*

commit a crime," he adds. "All the older guys that were here, it's *I'm going to do what I have been doing for years. And if I get caught that's going to suck, but this is what we do here."*

In the meantime, Lynne had begun working in the park. At first she held jobs at campground kiosks near Gold Bluffs Beach and Prairie Creek; later she worked maintenance at a nearby outdoor-education camp. Lynne was known around town as a friendly animal-lover: she often stopped by a local cafe to sip an espresso with the owner, her pet goose on a leash at her feet. She posted a photo of herself on Facebook, standing proudly in her park uniform. But things would sour between Lynne Netz and park management; one summer she was not asked back to work at a kiosk—because, she was informed, too many errors had appeared on her balance sheets. Feeling shunned, she grew a bit paranoid.

The discord was becoming generational. Like Garcia, Guffie, and Cook before him, Derek Hughes had developed a tense relationship with the Redwoods rangers in and around Orick. "They never come from Humboldt," he says. "A local knows how to treat a local. If they pull a local over, they know who they're dealing with. It eases the tension.

"I mimic how they treat me," he adds. "And they don't like it." He watched as RNSP chief ranger Stephen Troy and other rangers worked their way around the Blocks, trying to cultivate informers. One ranger in particular, Branden Pero, was Troy's "foot soldier," according to Hughes. Pulled over by the two rangers on a spot-check for timber, Hughes says, he turned to Pero and asked, "You really gonna fill his shoes?"

It was an absence that caught ranger Branden Pero's eye, snagging his attention from the road to the turnoff near May Creek;

the twinge of knowing that something was missing. Because his patrols normally include scanning the highway shoulders for signs of activity, Pero remembered that a pile of rocks normally sat to the left of the locked steel farm gate that separated Highway 101 from the creek. But on January 24, 2018, while on a standard patrol through Redwood National Park, he noticed that the stacked rocks had been scattered. Some had tumbled into a drainage ditch, while others were strewn around the gate itself.

About a quarter-mile up the road, Pero turned around and circled back. Parking his National Park Service work truck, he stepped out. To the left of the gate he saw tire tracks embedded in the earth where the rocks once sat. The impressions, spotty between the gate and the foliage, led to the middle of a small clearing along a path. Where the marks stopped, the ranger found a semicircle of sawdust and wood chips.

Pero pressed the Send button on the receiver pinned to the bulletproof vest he wears in the field, tilting his head slightly to the right to speak into it, and radioed ranger Seth Gainer back at the South Operations Center (SOC) about five miles south. "Looks like a cut site up here," Pero reported. As Gainer confirmed that he was en route to provide backup, it began to rain.

May Creek is part of a network of aquatic pathways that stitch through the redwood ecosystem in northern California, ultimately draining into the Pacific Ocean. Its shores are thick with brush that's difficult to cut a swath through, and it is not easy to reach or navigate. With his eyes to the ground, Pero followed what he identified as footprint and tire-tread indentations in the undergrowth. In some sections the brush had been flattened or crushed by something other than human feet, the shapes irregular.

Turning right, Pero noticed a *desire line*—not an official trail forged and maintained by park rangers, but one tramped down by a human's simple urge to walk there, wide enough for a lone hiker. Desire lines, especially in a rain forest, are faint and easy to miss, but they are everywhere. Pero could see one leading up the hill.

Suddenly, though, it began to pour. Pero hadn't brought a coat, so he ran to his truck and drove back to the SOC, where he changed into dry clothes and gathered some gear. He then drove back to the site, finding Gainer already working his way up the trail. The rain had stopped, but the air was still misty.

Camera in hand this time, Pero began to photograph the tracks. During ranger training he had learned to identify and assess the freshness of tire tracks in the woods, but it was growing up in the forest (Pero is from Redding, California) that had taught him to examine the tracks in depth. Over the years, he had learned to identify vehicle tracks when he set out to meet up with friends on back roads. At May Creek, he noted that the treads resembled those on Toyo brand tires: Pero had installed that brand on an old truck of his.

Pero joined Gainer on the desire line. Winding his way through about 75 yards of whipping branches and tall foliage to reach the top of a ridge, he turned to his left and spotted a redwood trunk whose base had been hollowed out. Through a cut more than three feet high he saw the flesh of the tree—pale wood in contrast to the outer bark. Sawdust that had drifted to rest under an overhang of moss and leaves was still dry and fresh.

Scattered about the cut site were clothing and equipment. Although Pero and Gainer could tell from channels in the wood that a chain saw had been used, a Fiskars ax also lay near the stump. So did a black work glove, presumably peeled off and discarded in the heat of the theft.

The ranger pair discerned a strategy to the cut: it had been hacked into the trunk on the side opposite the makeshift path, facing thick brush that would be almost impossible to penetrate, making it less likely to be found. The wood grain displayed the same bird's-eye swirl so prized by Danny Garcia five years earlier—a wavy, dark-amber pattern that flows like a disturbed pool of water. "Nice-looking wood for a table or whatever," Pero observed. Bird's eye is worth several times as much as more common hardwoods; it turns a rich red-brown when polished.

Pero began taking notes:

1. The stump was 30 feet in diameter.
2. It was not a fully intact tree. [This area of the Redwoods had been logged extensively before being sheltered within the national park. How remarkable, then, that the remaining stump was so huge— incapable of being fully framed in a photograph taken from the desire line.]
3. The redwood was still alive; new saplings had sprouted from its base. [Redwoods are unique for many reasons, notably that they are conifers able to grow from either seed or stump. They have a habit of popping up in massive "fairy rings" around the spot where an old-growth ancestor once stood, the new growth made possible by the life held within the knaggy burl that once protruded from the base.]
4. The poachers had been targeting the burl, not the trunk itself.
5. Items left at the site made it appear the thieves would return for the remaining wood at any minute.

Turning around to head downhill, Pero and Gainer caught sight of a nearby downed trunk that had been slashed to bits.

Cuts six feet long appeared to have been freshly made down its side, the wood slabs then peeled away.

Branden Pero joined Redwood National Park in the winter of 2016, after stints in Nevada's Great Basin National Park and the swamps of the Florida Everglades. His father worked maintenance for the National Park Service, and Pero grew up in the northern California logging city of Redding, surrounded by national forests to the west, north, and east.

Sometimes Pero's dad took him to work and let him roam the trails all day; he'd stop wherever he pleased to fish or study wildlife tracks. As he grew older, Pero knew he wanted to work outside, and he figured law enforcement would suit his preference for new challenges daily. But he didn't "want to be at work on my days off," he says, so initially he avoided the Forest Service. When Pero entered the National Park Service training program, he selected the career path designated P for *Protection*; that meant he would carry a weapon and enforce laws, as opposed to guiding visitors.

The job of a park ranger, especially in remote regions such as northern California, is often misunderstood as that of benevolent steward of pristine nature—all khaki uniforms and brown, broad-brimmed felt "flat-hats," reminding visitors to "leave no trace." In reality, the job has for decades included active law-enforcement officers—police in another uniform. That's not without cause: park rangers are more likely to be assaulted on the job than Border Patrol or even FBI agents.

A complicated calculus comes into play when a timber poacher is being pursued. Some rangers have been killed at work, including a ranger in Olympic National Park who was shot deep in the woods. In one study, researchers found that National

Forest officers were worried enough about their physical safety that they avoided going into the woods at all, or deliberately alerted poachers to their presence by revving their engines or following obvious patrol routes.

Pero's boss, Stephen Troy, had arrived at the Redwoods via Virginia's Shenandoah National Park and Philadelphia's Independence Hall. Troy was installed as acting chief ranger of RNSP just as the Danny Garcia investigation was winding down. (Garcia was sentenced on Troy's first day of work.) On the wall behind Troy's desk at the SOC hangs a framed illustration of a police officer handcuffing a caricatured criminal.

Eventually Pero made his way back to the Redwoods. He settled down with his wife and new son in the town of McKinleyville, 28 miles south of Orick. When we met, he enjoyed watching hunting videos with his young son, and they practiced elk calls together. The Redwoods had long been one of his professional ambitions. Pero knew he would come face-to-face with poaching after his transfer, and he was keen to work in a "high law-enforcement" park. Rangers in the Redwoods make more arrests than those in other national parks, and they regularly seize large amounts of drugs, intercept stolen weapons, or stumble across people carrying unlicensed firearms.

Over three years of rounds in the Redwoods, Pero became well acquainted with the park's unique geographic challenges: the major highway that runs right through its core; the tens of thousands of forested acres that rangers can't cover in a single shift; the fraying socioeconomic situation in Orick and other communities nearby. His charge is a resource made valuable for the same reason it requires protection: it's rare and beautiful and in demand. Orick's proximity to the park boundary means rangers must sometimes conduct searches or investigations on its streets or in its private homes or businesses; in so doing, they appear to take on the mantle of cop over ranger. When they visit the gas station or the post office after work, they are in uniform. It is not uncommon for citizens of Orick to complain that they feel monitored or targeted by rangers operating outside the park boundaries. The line distinguishing park rangers from police officers is often a blurry one—a ranger is not always an affable wilderness guide in a Smokey Bear hat, but not every ranger carries a gun—and Orick's residents are governed right at this margin.

In 2017, Lynne Netz's dog Mister died, and she fell into what she calls "yet another depression." She had found it difficult to leave Orick, though she wanted to. Yes, she had found community in the town, and always had someone to talk to. But she was also going through a separation from Larry. She was working at an educational center run by the park, but she had begun to feel that the park rangers judged her harshly because she was so friendly with the locals, even if, in her word, they "aren't the best of people."

By then her son Derek Hughes was living with his girl-friend, Sara, in a one-room shed that he had built behind Netz's bungalow. He had raised the roof on it, added a sleeping loft, and installed insulation and drywall. Hughes had his own woodstove to fill.

One day that summer, Netz and Hughes were walking her new puppy along Hidden Beach when two rangers approached and issued Netz a written warning because the dog was off-leash. Hughes tapped his cell-phone camera to life and filmed the encounter, then posted it on Facebook that night. In the video, he and Netz both insist the puppy was off-leash only so they could hoist it over a log on the beach. No matter, the rangers reply—it's not allowed. Compounding matters, they ticket Hughes for an improper truck registration.

The camera then points down toward the sand, and Hughes appears as an elongated shadow holding a fishing rod. "Am I still being investigated for poaching wood?" he asks, as the rangers request his mother's identification and write up the ticket.

"No," a ranger responds.

"Danny got out of jail," we hear Hughes tell his mother on the video. "I read his discovery and apparently I was under investigation for falling a redwood tree. Blew my mind!"

Soon after that episode, Lynne Netz was fired from her job at the park's educational center. She links her termination to the run-in on the beach.

Chapter 14

PUZZLE PIECES

"[The park is] not for the locals."

—Chris Guffie

Standing at the base of the hollowed-out redwood stump, ranger Branden Pero devised a plan. He and colleague Seth Gainer hid two small motion-sensing cameras in trees near the locked steel farm gate back on Highway 101, burrowing them into the foliage so only their lenses peeped through. Then, retracing the path toward the poaching site, they installed six more.

The cameras that Pero and Gainer had cached in the trees would need to be collected regularly so their footage could be downloaded for review back at the office. On a rainy February day in 2018, Pero traveled back to the May Creek site and switched out the cameras' SD cards for just that purpose. While there, he discovered fresh vehicle tracks on the creek landing: yet more wood had been cut from both the downed log and the redwood stump. Not only that, but some of the wood blocks that had been resting near the stump had now been removed.

Back at the SOC, Pero sat at his desk in a small, windowless office. On the wall hung a John Wayne poster bearing the exhortation COURAGE IS BEING SCARED, BUT SADDLING UP ANYWAY.

Pero uploaded the footage from the SD cards to his hard drive. Because the cameras capture images using infrared, the photos appear mostly black-and-white. They are also very sensitive to even slight winds, so the first few pictures he examined were duds. Eventually, however, photos with a time stamp of February 2 revealed a small, light-colored truck arriving and making a three-point turn in the sword fern. Part of the camera's view was obscured by a branch in front of the lens, but in another photo Pero could make out the blurred outline of the truck's driver: he was standing by an open door smoking a cigarette, its lit tip flaring.

Pero tried to zoom in and enhance the image. No luck. Even so, Pero thought he recognized the man's stature and body type: tall and very thin. He suspected it was Derek Hughes, a man he had seen in town who regularly took wood off the beach. Reviewing additional photos, Pero saw the same truck appear multiple times, though the camera did not always capture its occupants. He walked to chief ranger Stephen Troy's office.

"Who does this look like to you?" he asked, showing him the photos on his laptop.

"That looks like Derek Hughes," said Troy.

"I think so too," Pero replied.

———

Troy remembers his former boss, Laura Denny, once saying that even though they always suspected he was involved in timber poaching, they "never could get Derek." He still can't prove it, but Troy suspects the amount of wood taken from the park by Hughes is "staggering."

As the SOC rangers began to narrow down the suspects in the May Creek case, they visited burl shops and drove around Orick to talk to "basically anyone who would say anything," says Troy. "A lot of people don't like to come to [the SOC]. They don't want anyone seeing them coming in here," he says. So he approached people on the street, knocked on doors, visited local businesses. He and Pero also drove by Lynne Netz's house, where they examined the tires on a small gray Toyota pickup truck and identified them as Toyo.

Those conversations, paired with the trail-camera evidence and the tire analysis, all pointed toward Hughes. Nonetheless, it took three months for Redwoods rangers to secure a warrant from the Humboldt County district attorney to search Hughes's property. Still, they considered themselves lucky—it had taken nearly double that long to gather enough information to nab Danny Garcia.

One challenge that law enforcement faces when it comes to prosecuting timber poaching is securing arrest warrants and moving cases forward in the court system. Many district attorneys don't want to take on poaching cases because, according to Denny, "when there's murders and there's rapes or whatever going on, that's going to take higher priority than someone taking a tree." The punishments are often middling, and Humboldt's prisons are overcrowded. The local legal system therefore has a high expectation of concrete evidence that will justify prosecuting a poacher.

At the time that Hughes's case was being investigated, a new deputy district attorney arrived in Humboldt and began making a name for himself by prosecuting environmental crimes. Adrian Kamada is known for his commitment to wildlife prosecution, and he started the job fully aware of the wood-poaching cases that plagued the park. Kamada had a strong interest in environmental crime, and he was keen to support law enforcement working to prevent it. As Hughes was being caught on forest cameras, Kamada was prosecuting one of the most bizarre environmental crimes in recent Humboldt history: poachers had scaled the cliffs near Orick, pilfering thousands of rare *Dudleya* succulents (commonly known as "liveforevers") to put up for sale online and in overseas markets.

Based on the clues they had gathered, the Redwoods team began working to secure a search warrant for Netz's property. This meant not only petitioning the court for the warrant but arranging for multiple law-enforcement agencies to provide support.

As chief ranger Stephen Troy tells it, executing a search warrant always starts with a bit of fear: "You just never know how people are going to react when you knock their door down," he says. "It's exciting in one way, nervousness in another. They are dynamic, for those first 15 minutes."

On the night of March 27, 2018, rangers from both the national and state parks met to review the operational plan for executing the search warrant that evening. Hughes was at home with his girlfriend, Sara. The back bunkhouse, which Hughes calls his "shed," had become a private enclave for the couple, a place apart from Lynne and Larry's house. Inside, a sheet hung between the front door and the rest of the room, blocking bugs from entering. Straight ahead from the front door, a ladder led up to a loft Hughes had constructed; to the left was a living room and bedroom, with a TV and coffee table.

Soon Pero, Troy, and a team of law-enforcement agents approached Lynne's house and knocked on the door. Pero announced that they had a search warrant, and the three people inside—Lynne; Hughes's sister, Laura; and Larry—came out and stood on the front lawn as the team began its search.

Circling the yard, Troy and Pero approached the shed. Hughes and Sara were lounging in bed watching TV, so Hughes was surprised when the barrel of an AR-15 semiautomatic rifle suddenly swept the bug sheet aside and came to a halt, trained on his face. He recognized the officer carrying the weapon as Troy—and noticed that the ranger's finger was resting on the trigger.

"Sir, get your finger off that trigger," Hughes remembers saying. "If you shoot me, you're going to have all hell to pay."

And he remembers Troy's response: "Shut up and get out of here."

Outside, Pero pushed Hughes to the ground, handcuffed him, and pressed his head against the rim of a truck's tire. Troy remembers Hughes as being "really mouthy—mainly with me, which is fine." Hughes denied having any involvement in poaching wood but according to court documents he turned to ranger Emily Christian and allowed: "Yeah, I have meth." He also claimed that he got the wood found lying around the property from a mill-worker friend who had dug it out of the scrap pile.

The five family members were then left standing in the backyard while park rangers swept the shed.

"And boy," says Troy, "we found a lot."

Inside they discovered what appeared to be brass knuckles (though Hughes says it was merely a belt buckle shaped to resemble those illegal items), a small plastic bag containing traces of methamphetamine, and four meth pipes. On a shelf sat a handgun—and a set of old keys that had gone missing from park facilities. The search team also seized a laptop and Hughes's cell phone.

As the rangers prepared to leave the shed, Pero noticed multiple sheets of paper thumbtacked to a wall near the door. They proved to be court documents from Danny Garcia's tree-poaching case—as Hughes had mentioned to his mother during their run-in with the rangers on Hidden Beach, he truly *had* read the discovery.

Outside, the parks team searched an RV parked in the driveway and found a motion-sensor camera engraved with the initials *REDW*—a device the park would normally hide in foliage. The search team also came across stacks of chopped redwood in three locations on the property: alongside a boundary fence; hidden beneath a tarp on Netz's deck; and inside a woodworking shop attached to the garage, which also held a wood lathe and redwood chunks in various stages of being turned into bowls.

"I was taken aback to see he was turning it on his own," says Troy. "Our experience in the other cases was that they sold it raw, in slabs." But a few larger slabs had not yet been turned. "We could tell [those slabs were] going to slide right into the hole we're investigating," says Troy. "If he had turned it all into bowls, we would have been very hard-pressed to say it was from that [May Creek] tree."

Hughes was loaded into the back of a patrol truck and driven to the Humboldt County Superior Court building in Eureka, where a clerk booked him on six criminal charges: felony vandalism, receiving stolen property, grand theft, possession of metal knuckles, possession of methamphetamine, and possession of methamphetamine paraphernalia. He left the building and waited for Netz to come pick him up.

"At the end of the day," Derek Hughes told me later, "the [May Creek] wood was at my house."

The National Park Service struggles to prosecute poachers. California's parks are vast, and park wardens face an impossible hurdle: their charges cannot speak for themselves, cannot serve as their own witnesses. So it becomes a game of chance—you pray for a visitor to notice a cut and report it, or you hope to catch a poacher in the act.

In 2014, amid the rash of burl cases being investigated by rangers Rosie White and Laura Denny, Redwood National and State Parks made the decision to close all roadside turnouts and parking lots along Highway 101, in essence instating a ban that would make any parked car stick out like a sore thumb. The problem, says Pero, was that some poachers preferred it that way: they could hike in and cut when the road was shut down. Then, with no one passing through the area either by car or on foot, they could stash the wood behind a tree and return the next morning to cart the load out. "All the hard work would be done," Pero explained.

When park rangers such as Pero call their options for catching poachers *limited,* they mean so on both a logistical and financial scale. To remotely monitor every at-risk old-growth tree in Redwood National and State Parks alone would be utterly impractical. The same goes for cordoning off the park with fences to keep out night-hikers or those who drive in at midnight.

Instead the Park Service has installed a tip line, but Pero calls the occasional messages left there "just weird." So rangers rely on serendipity, bolstered by regional contacts—the spark of recognition of a cut site on a field-day patrol, sometimes followed up on with local informants who might know something.

"Realistically," says Pero, "[we could dedicate] one or two full-time rangers specifically to try and find this stuff. But finding the money to fill positions solely for that, when we have other things that tie us up..." Still, personal grudges and turf wars in the woods have prompted poachers to sell each other out to law enforcement. In some cases, the Park Service has offered to forgive minor charges pending against a confidential informant if they furnish information on a poaching site. In the same vein, the Redwoods Park Association has partnered with the Save the

Redwoods League to offer a $5,000 reward for anyone pointing them in the direction of poachers.

—————

Chief ranger Troy was relieved that executing the search warrant had been fruitful, but he also knew the steps required to convict Hughes were incomplete. On May 9, 2018, half a dozen park employees loaded up some of the wood seized from Netz's property and drove it back to the May Creek site. The group placed one large stump and three slabs of redwood in a motorized wheelbarrow, then pushed it up the desire line. They turned the samples this way and that atop the stump. The wood blocks slid together like perfect puzzle pieces.

As he investigated the poaching case against Derek Hughes into June 2018, ranger Pero approached a Netz neighbor by the name of Robert Anderson, who happened to be incarcerated at the time in the Humboldt County Correctional Facility. During their conversation, Anderson acknowledged that Hughes was a friend. Very late one night, Anderson recalled, Hughes had approached him with a request to "help him with something."

On two separate occasions in the winter of 2018, Anderson divulged, the two friends had taken a 10-minute nighttime drive from Orick to a redwood stump. Anderson described the approximate location: a forested landing near a bypass. Hughes, he said, had brought a chain saw.

Anderson recounted walking "around the woods in the middle of the night, looking for a specific tree." On their second trip two nights later, they found the tree and began to gather up the redwood chunks surrounding it, rolling them down the hill toward the truck. Anderson mimed the size of the blocks with his hands: only yea big—24 inches or so. He'd be making bowls from the wood, Hughes told Anderson.

The investigation would be bolstered by a recording of a conversation between Anderson and Hughes that took place on the morning of May 24 over the phones at the county jail, which had been recorded by a district attorney's office investigator. While Hughes was visiting Anderson in prison, the two discussed a recent newspaper story about Hughes's case. Thanks to the news item, gripes a voice the Park Service has identified as belonging to Hughes, he would be unable to get a job: "It's almost like they *want* me to cut redwood burl." The voice went on: "In the article, they made it sound like we cut down a redwood tree. It was a *stump*."

Anderson had supplied Pero another useful tip: he should follow up with someone named Charlie.

Pero knew Charles Voight from around town. Voight had been raised in Orick and worked at a mill with Larry Netz for years. Voight and Hughes were friends, and they were often seen together. At the beginning of the May Creek poaching investigation, Pero had pulled Voight over and found a meth pipe on him. Rather than prosecuting the violation, the park resolved to leverage the opportunity: they would drop the drug charge in exchange for information on the Hughes case.

Voight worked at a market located across Highway 101 from the SOC, so Pero and ranger Emily Christian walked across the street one day and asked Charlie Voight to drop by the operations center when his shift was done.

That evening, according to Pero, Voight described visiting a stump in the park with Hughes at the latter's invitation. Voight insisted that he had not known what Hughes was planning beforehand, but once they reached the site he witnessed him cutting blocks out of the stump with a chain saw. He described the location: about 10 minutes north of town, near a bypass, off the highway. Voight himself had refrained from any cutting, he claimed; rather, he had been a lookout, and had helped load the wood into the back of a truck.

Pero felt certain that Voight—a bespectacled five-foot-nine—was not the man captured in the surveillance photos. (The unidentified suspect's facial structure seemed not to match Anderson's, either.) Hughes was the common denominator, although he later claimed that Voight had borrowed his truck, only to return it laden with wood. Hughes then offered the wood for sale to a flute maker, but the artisan declined to pay the high price asked.

The cameras remained hidden in the trees at May Creek. "We let [Hughes] know that we were watching," Pero explains. "If we saw him driving through [the park], we'd pull him over—he always had some kind of vehicle-code violation—to see what he was up to."

Later that summer, the rangers received a report detailing communiqués found on Hughes's phone. Numerous text messages—some with photos—concerned the sale of old-growth redwood slabs.

Hughes says that he would never do what Garcia did: hack into living trees. "If it's on the ground, it's already dead and on the ground. We're not going around raping these trees."

At a pretrial hearing, rangers and lawyers for the National Park Service gathered along with Hughes and his defense team in a courtroom at the county courthouse in Eureka. Hughes looked on as the Park Service presented its evidence in the case. For years he had maintained his innocence during research and consultation with lawyers and the Park Service.

On sheets of paper, the park rangers identified images taken by Pero's hidden cameras, showing the man in the headlamp glow. "None of them would even look at me the whole time they're on the stand," Hughes later recalled.

The prosecution also called to the stand Mark Andre, the Arcata forestry expert who had estimated Danny Garcia's depredations at $35,000. After the wood in the Hughes case had been seized from Netz's property, Andre was asked by Pero to assess its value too. Now, returning to the SOC evidence locker four years on, Andre photographed the 32 pieces of wood he found stacked there (he would assess the grain later at a computer). Then he visited May Creek with Pero, where Andre took "measurements of the void" on both the trunk and the log using a tape measure and a Biltmore stick. He made some notes in his field journal, then went back to his office and began the process of calculating the wood's market value. Andre used his measurements to convert into board feet the amount of wood removed from the site. The grand total: 285 board feet of poached wood. Finally, Andre used that number to consult with local mills and calculate the price the wood was likely to fetch at retail.

"[I] had some buyer information for that specialty market," Andre observed in court, "which is really, you know, a small market. The few buyers don't always like to talk about it too much, but the ones I talked to swore they only purchase from legal, permitted sites."

Board feet is the standard metric for assessing timber value, but some buyers pay by weight. (The haul in the SOC evidence room came in at 1,330 pounds.) Other buyers prefer to pay a flat rate per piece: in that case, Andre conservatively estimated, the poached wood could go for about $50 per chunk. He was convinced, however, that board feet would yield a better calculation of the wood's actual worth. In the end, he valued the wood at $625.50.

Clearly, the precise value of old-growth redwood is difficult to assess. Because it constitutes less than one percent of all lumber sold annually in the United States, its rarity insulates its value from such broader market whims as interest rates, home remodeling, and housing starts. But the *true* value of a redwood must factor in its impact on biodiversity, its influence in the forest, its power for tourism, and its place in our culture.

"By the way," Andre told the courtroom in concluding his testimony. "The stump is still alive. It is sprouting. It still has second-growth redwood sprouts coming off of it. It is a stump, but it is a living tree."

Four women pose by a "Save the Redwoods" banner displayed on an auto. (*Peter Palmquist Collection of Humboldt County, California, Male Photographers, Yale Collection of Western Americana, Beinecke Rare Book and Manuscript Library*)

The first "Save the Redwoods" gathering (*Save the Redwoods League photograph collection, BANC PIC 2006.030—B, The Bancroft Library, University of California, Berkeley*)

Among the redwoods in California
(*1999.02.0337, Ericson Collection,
Humboldt State University Library*)

Traditional Yurok house
(*2003.01.3304, Palmquist
Collection, Humboldt
State University Library*)

Redwood with burl near road
(*2003.01.1731, Palmquist Collection,
Humboldt State University Library*)

Judi Bari (left) and Darryl Cherney at a Redwood Summer protest, 1990 (*courtesy of* Ukiah [CA] Daily Journal)

The "peanut," trucked by loggers to Washington, DC, in 1977 as part of the "Talk to America" convoy, outside Orick's Shoreline Deli and Market (*David A. Bowman,* American Field Trip)

A calendar commemorating the convoy (*David A. Bowman, American Field Trip*)

A chain-saw slash left in the body of a redwood in Redwood National and State Parks (*David A. Bowman,* American Field Trip)

A partially cut log rests on Hidden Beach, outside Orick. (*Lyndsie Bourgon*)

The poaching site discovered by Preston Taylor near Redwood Creek (*courtesy of the National Park Service*)

A Redwood National and State Parks ranger measures the cut site found at May Creek. (*courtesy of the National Park Service*)

A camera hidden by National Park Service rangers in the foliage of a redwood (*Balazs Gardi*)

Redwood National and State Parks ranger Branden Pero on a beach near Orick (*Balazs Gardi*)

A slice of redwood burl held in storage at an Orick burl shop *(Balazs Gardi)*

A redwood burl bowl turned by
Derek Hughes (*Derek Hughes*)

Derek Hughes's living
quarters in Orick
(*Derek Hughes*)

Danny Garcia and his daughter (*Jennifer Paddock, courtesy of Danny Garcia*)

Poached lumber ready for transport on the shores of Tres Chimbadas Lake, Peru (*Lyndsie Bourgon*)

One of four shihuahuaco poaching sites in the Infierno conservation concession, Peru (*Lyndsie Bourgon*)

The stump of a poached shihuahuaco covered by dried palm fronds (*Lyndsie Bourgon*)

A forest guardian monitoring Comunidad El Naranjal land, Peru (*Lyndsie Bourgon*)

Illegally logged land converted to agricultural use outside Comunidad El Naranjal (*Lyndsie Bourgon*)

Ruhiler Aguirre on Tres Chimbadas Lake, Peru (*Lyndsie Bourgon*)

Wood samples from xylaria collections to be added to the USFWS lab wood database (*Lyndsie Bourgon*)

A chemist from Ed Espinoza's Timber Tracking team at the US Fish & Wildlife Service Forensics lab feeds a sliver of wood into the mass spectrometry DART machine. (*Lyndsie Bourgon*)

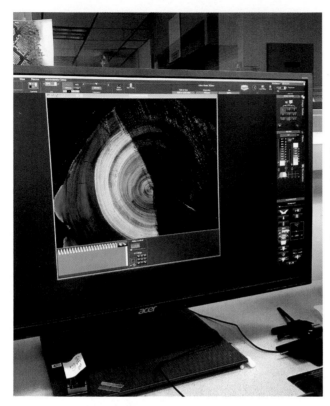

A cross section of black locust wood is analyzed using fluorescence at the USFWS lab in Ashland, Oregon. (*Lyndsie Bourgon*)

Justin Andrew Wilke, involved in a poaching case that led to a massive wildfire, atop a pile of maple logs. (*court exhibit, United States Attorney's Office, Western District of Washington*)

A poaching site in the Sunshine Coast Community Forest, British Columbia
(*Veronica Alice*)

Bark shaved into shingles in the Sunshine Coast Community Forest *(Veronica Alice)*

Chapter 15

A NEW SURGE

"I guess you could say people are just trying to make a living, that's all there is to it."

—Chris Guffie

As Derek Hughes waited for his trial to proceed in 2018, the Pacific Northwest experienced another increase in timber poaching. Timber prices were running hot for several North American species, including Douglas fir. Record-high prices of more than $440 per thousand board feet—that's 143 two-by-fours, or roughly two cords of wood—were reached in February 2018. With the payoff for poaching wood so high, some found it increasingly worth the risk.

In Washington state, the term *epidemic* was applied to the spread of poaching wood from public land. "There are desperate people," read a post on the Washington Forest Protection Association blog at the end of the year. "Some of them addicts, who are willing to meet that demand by illegally cutting wood from federal and DNR [the state's Department of Natural Resources] land."

So much poaching was taking place on Vancouver Island

in 2019 that Natural Resource officers (NROs) were "run off their feet" by the number of trees being felled on their land. British Columbia's NROs are charged with preventing revenue loss from provincial forests. They enforce laws and regulations, working in all-black uniforms that include bulletproof vests. The rising rates of forest crime prompted the province to train its NROs in hand-to-hand combat, and a report recommended that they carry batons and pepper spray.

One sunny spring afternoon in 2019, I shadowed Luke Clarke, an NRO based out of the city of Nanaimo, on his patrol rounds. In the course of just a few hours together, we stumbled across multiple poaching sites along a stretch of road scarcely a mile from the central highway connecting Vancouver Island's cities and towns.

The shoulders of British Columbia's forest-service roads are dotted with stumps where there should be towering trees. The provincial forests are still rich with coastal Douglas fir, big-leaf maple, hemlock, and groves of old-growth cedar. Over time, residents and environmental activists have persuaded the province to limit logging in many areas near cities and towns. But official restrictions have failed to protect these places from an unprecedented rash of timber poaching. As with redwood trees in northern California, lofty roadside trees made ideal targets because of the very curtain they provide, and in this case an entire tree could be quickly bucked up and loaded into a vehicle. "I honestly can't keep up," Clarke said. "There's just so much of it." From 2013 to 2018, NROs in British Columbia reported some 2,300 forest crimes. The most common were timber theft, illegal harvesting, and arson.

The trees targeted during that time were mainly Douglas fir, cedar, and maple, their wood converted into everything from lumber to shake blocks.

Douglas fir was most commonly chopped up and sold for firewood—it burns hotter than other species. Wood could be found for sale on roadsides, or posted online on platforms such as Facebook Marketplace, priced at a flat rate for a truck bed.

Cedar usually ends up as shingles, shake blocks, or furniture, not firewood. It is roughly three times more valuable than Douglas fir. Cedar's attributes—its deep, rich color; its woodsy scent; its habit of stick-straight growth—keep it valued and sought after, especially by builders of saunas and decks.

Maple, notably, requires a certain skill to fell. Unlike a redwood or a Douglas fir, a maple tree branches off, forks, leans, and tilts. The historic maple market was primarily sawmills that turned the trees into music wood, but in recent years Clarke had witnessed a new demand surface: "You go to almost any bar in Vancouver, they have that maple, live-edge board table," he said. "It's such a common thing. Where is it coming from? Some of it's legitimate, some of it's not."

If you know what to look for, the poaching in British Columbia is often easy to spot: muddy tire tracks leading onto a highway from a forest lane strewn with pine needles, with tree branches left scattered along the shoulder. "It's like taking one cookie from a jar—no one notices," said Corporal Pamela Vinh, one of two investigators with the Royal Canadian Mounted Police (RCMP) Forest Crime Investigation Unit, which is based in British Columbia. "But a bunch in a row? It's more noticeable."

Clarke and I had been driving toward a previously identified poaching site, but new sites were popping up so quickly that he wasn't surprised when we came across a fresh one. (In the course of a scant hour, we passed three poaching sites in stands of old-growth Douglas fir.) Clarke's job had become monitoring the gray, fluted, dead-standing Douglas firs that provide homes to some of the region's endangered species.

At the new site, Clarke hopped out of the truck and began to unload the tools he needed to begin an investigation. Using a measuring tape and planting numbered plastic flags, he measured the tree and photographed its remaining stump and trunk. With a tablet computer, he then punched this information into a database that would be used to track further investigation of the site. Finally, Clarke scanned the ground for tire tracks or boot prints.

The poached tree had stood in a forest parcel now home to tent communities of unhoused people. In one spot, a deep thicket of woods was intersected by a creek, and the ground around the area was crowded with dead-and-down trees. Through them we spied a makeshift wood bridge that had been placed over the creek, making it easier to cross. The location had been part encampment, part wood-poaching operation—wood had been transported over that bridge and sold as firewood along the roadside or to clients in the nearby city of Nanaimo.

The city of Nanaimo has much in common with communities in the American Pacific Northwest that struggle with a shifting economy and waves of chronic unemployment. For many poachers on Vancouver Island, market demand is only part of the motivation: British Columbia faces the intertwined crises of homelessness and opioid addiction. In 2018, Nanaimo was home to Canada's largest tent city (dubbed *DisconTent City* by its estimated 300 residents). The city's per-capita homelessness rate is among the highest in the province.

The tent city created a crisis of identity in Nanaimo, and citizens flooded radio shows and newspaper comment sections with their opinions. Whereas many locals would prefer to believe that the homeless are outsiders, a November 2018 presentation from the provincial agency BC Housing indicated that the majority of people in the tent city had lived in Nanaimo for years.

Next door to the NRO offices is a brand-new housing facility, built to accommodate some of the denizens of DisconTent City. In its parking lot, Clarke occasionally encountered a man he had previously charged with timber theft; it turned out he was living in the facility. "He's pretty hard up, and is honest and transparent with me when I ask him questions," Clarke said. "He says you make money just selling firewood."

Clarke had posted signs high up on electric poles at the entrance to various trails and roads, publicizing an anonymous tip line to report timber theft. He and his fellow officers installed hidden cameras in trees along the highway that cuts through town. He spent time seeking out informants from the town around him.

In British Columbia, this is often where cases stall out in the system. Much like their American counterparts, Canadian conservation officers must formulate airtight cases against poachers in order for them to be successfully tried in court. Only half of the province's 2,350 instances of forest crime tracked between 2013 and 2018 were investigated and prosecuted; of those, a mere 140 made it to the courts.

A further challenge to stemming timber theft is applying the appropriate penalties to those found poaching. How can we place a numerical value on the deeply holistic worth of an old-growth tree?

If anything, things got even worse in British Columbia in 2020 and 2021. One spring day in April 2021, I received an email from Sara Zieleman, an administrator at the province's Sunshine Coast Community Forest (SCCF). Zieleman was writing to let me know that a 200-year-old Douglas fir nearby had recently been poached. "In our area," she noted, "old-growth has a significant deficit. So it's a real loss."

A pocket of British Columbia's mainland southwest, the Sunshine Coast is a 110-mile stretch of road similar to the Redwood Highway—a winding strip of pavement surrounded by a vast forest dripping with moss; bound by inlets, fjords, sounds; nestled under the Coast Mountains. No redwoods grow here, but the region's lush biodiversity features towering exemplars of cedar, Douglas fir, and hemlock. A network of craggy islands dots the shore, accessible by ferries or private boats. Some of these isles are small, and home to no one at all. Others are speckled with summer houses. Those who live here full-time often are well-versed in the region's rich logging history.

On the highway just north of the town of Gibsons, the Sunshine Coast Community Forest is a remnant of that era. Founded in 2003 in the shadow of British Columbia's War in the Woods, the community forest is managed for conservation, recreation, and timber harvest by a volunteer board. The outfit aims to step into the void between managed conservation sites and commercial use, offering community firewood salvage and harvest through free permits. The SCCF has developed harvest plans and maintains its forest for the financial return. But it also takes into account watershed management, sustainable road development, and untouched reserves for wildlife.

"We'll hand out hundreds of permits a year to people," SCCF operations manager Dave Lasser told me in 2020. "They've all got woodstoves—they love it." He has been gathering his own firewood from the same precinct of the community forest for four years. "I'll get 10 to 12 loads this fall," Lasser reckons, "and I can burn it all in my woodstove. Getting firewood is good therapy."

Still, the community forest is beleaguered by poaching. In the spring of 2019, around the same time Luke Clarke was tracking poachers on Crown land, Lasser declared that the forest was suffering an "epidemic" of poaching. Over the previous five years, he estimated, some 1,000 trees had been poached from the community forest and Crown land directly surrounding it. Lasser has counted hundreds of stumps from illegally felled trees along a single stretch of forest road. One cluster of mixed fir, hemlock, and cedar stand has gradually become just a hemlock-cedar stand, its fir trees lost to firewood thieves.

Sometimes a member of the public will call the SCCF office to let them know they've spotted a felled tree out on the property. At other times Lasser has come across the trunk of a downed tree near the road on one of his regular drives through the SCCF's blocks. Once, he found a felled 70-foot Douglas fir simply abandoned on the ground, the surrounding forest impenetrable to any truck.

"More often than not what people will do is find a tree, fell it across the road perpendicular, then buck the guts out of it, from ditch to ditch," says Lasser. "Pop it into rounds, throw it into a truck, and gone." In many cases a tree's extremities—its crown and the trunk segment nearest the stump—are discarded beside the road amid a pile of sawdust. "Sometimes you go up roads that haven't been used in a while," he says, "and you'll find somebody's started dumping trees like [it's] their own private little firewood forest—you know, 10 trees in a 50-meter-square area."

One day in May 2020, Lasser drove past a towering Douglas fir with a large notch cut from its trunk, a car jack wedged irretrievably in the wood. He suspected that poachers had attempted to fell the tree using a chain saw too puny for the task, then abandoned the project with the jack still stuck in the trunk. "It was a real mess," he laments.

After its discovery, Lasser installed a surveillance camera in the branches of a nearby tree, hoping to catch the poacher returning for more wood. A few days later that camera captured a man, chain saw in hand, felling the notched tree perpendicular to the road. The poacher then methodically bucked it into rounds and loaded the forbidden wood onto a trailer attached to the back of a pickup truck.

But it was 2020, and some members of the province's compliance-and-enforcement team had been reassigned to patrol the US border (closed to nonessential travel since March in response to the Covid-19 pandemic). Tree planting and logging had been deemed necessary services, so lots of forest-compliance officers had been detailed to monitor those activities to ensure they followed COVID-19 protocols. Poaching fell far down on the list of priorities.

Yet the demand for wood was only growing. A truckload of wood was fetching an unprecedented $300, and anyone with a spot in mind could harvest a couple of truck beds a day. (Farther south, closer to Vancouver and the province's populous southern mainland region, that same truckload commanded a staggering $800.) Lasser had even heard whispers that certain drug dealers had begun requesting payment in wood, not cash; they could truck the plunder a couple of hours south and sell it for a quick profit.

The link to drug dependency is obvious, in Dave Lasser's eyes. "It's because they need the cash for the…habits they have," he told me warily. Yes, the SCCF's popular free-firewood program allows anyone to take as much as they want, but a "personal-use" harvest permit must be granted by forest administrators, and gathering can occur only outside of fire season, "typically between fall and late spring." This restriction outlaws harvesting from April through October—prime time for tourists to sit around a campfire.

A tree must be at least 250 years old to be granted old-growth (or "class 9") status by the province of British Columbia. The

Douglas fir taken from the SCCF had been left standing in the express hope it would reach that milestone. Forest staffers look for trees displaying old-growth characteristics—reaching more than 100 feet tall, growing in lush stands—and deliberately leave them intact. Often you will see what Lasser describes as "single stems scattered through the stand" of logged wood, looking stark and lonely amid a felled plot but slowly aging into old-growth. They become homes for raptors, which perch in the branches and crown and wait for prey to venture into the cut block, then swoop down to snare a vole or a field mouse. The fir pierced by the car jack had a damaged crown (broken off in a windstorm), but it was spared further molestation as a "wildlife tree" in which animals could nest.

As the price of a cord of wood has soared, the risks associated with poaching have become increasingly worth bearing for perpetrators. "If a guy can go out and get two pickup loads a day, he's got $500," Lasser says. "Even if you convict someone, what is the fine—$180? That's not even the price of a cord." Instead, many NROs and RCMP officers focus on traffic stops, staying vigilant for truck beds bearing wood and asking to see the driver's personal-use firewood permit. If no permit can be produced, the wood is confiscated and the driver fined.

Properly protecting old-growth wood, Lasser believes, will require not only steeper fines but an increase in the number of forest wardens operating deep in the wilderness. *But might that not simply displace the crime,* he wonders, *driving poachers to earn money elsewhere?*

"Your house," asks Dave Lasser, "or the bush?"

Chapter 16

THE ORIGIN TREE

"There are those people who trust certain people and it's a little circle, I guess you could say. You have people who still have honor amongst men."

—Chris Guffie

A plume of smoke rose above the Olympic National Forest on the morning of August 4, 2018. August is the thick of fire season in the Pacific Northwest, and the Forest Service's wildland firefighters were quick to track down the source of the smoke, which had been reported to them by a hiker in the Elk Lake Lower Trailhead area.

A group of three firefighters tracked the path of the smoke to a ravine close to a popular hiking trail. The fire had started near the banks of Jefferson Creek, surrounding the base of a mature maple. The maple would soon be dubbed the "origin tree"—the point from which a 3,300-acre forest fire would spread.

At the site, a forest fire was burning low. Firefighter Ben Dean studied the ground and determined that the fire likely started in a hollow at the base of the tree, but it was unclear what had sparked the fire; the area was particularly humid, the forest floor damp. Dean noticed the site looked like someone had prepared it for logging: a nearby Douglas fir had been trimmed to facilitate access to the maple, and a check mark had been sprayed on the maple's trunk. Near the bottom of the maple sat two cans of wasp-killer spray. Not far away Dean discovered a red canister of gasoline, as well as a camouflage-print backpack containing some classic tools of the logger's trade: chain-saw chain, locks, wedges, and oil.

While the three-man firefighting team was making notes at the site, another Forest Service officer, David Jacus, arrived to help. As he neared the location, Jacus was passed by a local man whom he knew to be Justin Wilke, driving a white Chevy Blazer. Meanwhile the fire had gained momentum, with flames climbing the trunk of the maple and starting to lick the canopy. The firefighters felt the heat of the fire and knew they didn't have the proper equipment to quench it. Dean felt it was too dangerous for the team to stay there much longer, so they gathered what they could of the evidence—the backpack and the red gas can—and left the scene.

Dean decided to spend the night in his truck at the trailhead. He handed the evidence over to Jacus and settled in to monitor the fire's spread from his vehicle. Jacus knew that Wilke had been staying at a campground about 150 yards from the origin tree, so he drove to the site and found Wilke in a white camper trailer. Wilke denied having poached maple in the area. "I don't even have a chain saw," he said. That night, however, a white vehicle came barreling toward the trailhead where Dean slept in his vehicle. The white vehicle slowed as it approached Dean's truck (emblazoned with the distinctive green-and-gold Forest Service shield), then mysteriously turned around and left.

The Forest Service was unable to contain the fire, which would gain notoriety as "the Maple Fire." It burned for more than three months—until November of that year—scorching both National Forest and state land and ultimately costing $4 million to contain. Once the Maple Fire had finally been extinguished, the Service brought in a specialist, who confirmed the start of the blaze as the base of the origin tree. Due to the humidity, the specialist noted, it was likely that an accelerant had been used to get the fire going. On further inspection, the Forest Service found three additional maple-poaching sites in the area, their telltale stumps clumsily concealed beneath branches and debris. The total value of the felled trees was estimated at $31,860.

Though Justin Wilke had been camping in the area that month, Forest Service rangers knew he also often stayed in another trailer, at a property about nine miles from the Elk Lake trailhead. In his subsequent investigation, Jacus visited that property and found wood shavings and small blocks of maple sitting near the trailer in the yard—which, stacked with discarded machinery, tools, and furniture, bore a striking resemblance to the Cook Compound.

The owner of the property, Alan Richert, confirmed that the backpack found at the origin-tree site was indeed Wilke's, and that Wilke had been poaching maple in the forest that summer with another man, named Shawn Williams. The pair would bring the illicit lumber to Richert's yard for processing into blocks. (They had found a mill in nearby Tumwater that was willing to buy the wood.)

The Maple Fire case is a prime example of the leverage that informants can exert in timber-poaching episodes. In the course of the Forest Service's investigation into the Maple Fire, numerous people in Wilke's circle sold him out. One friend told investigators that Wilke had identified the origin tree for harvest, but that he had been deterred from felling it by a wasp's nest in its branches.

Wilke then floated a disastrous idea: *What if they just burned the wasps right out?*

The next day, Wilke and three others made their way to the origin tree, poured gasoline on the trunk, and set it aflame. At first they thought they had the fire under control. When the flames started to spread uncontrollably, however, they tried to extinguish them using Gatorade bottles filled from a nearby creek. The conflagration sent the group scattering, and Shawn Williams eventually got a ride home from a friend, who later testified that Williams had complained of being stung on the hand.

The next morning, there was smoke in the air. Wilke was agitated and hid a chain saw.

The Forest Service officers continued to press their investigation. The owner of the Tumwater mill showed them a purchase ledger documenting that he had bought maple from Wilke no fewer than 22 times over a five-month span; in his back room stood a stack of hundreds of blocks reaching halfway to the ceiling. During every transaction, Wilke had displayed a permit attesting that the wood came from private land. When Forest Service officers visited the property in question, however, they found nary a maple stump—nor even a single maple tree.

The three maple stumps unearthed in the forest, though, would soon make history. The Wilke case would become known not only for the massive wildfire it unleashed but for the novel means deployed to bring its instigator to justice: his was the first trial to use tree DNA in a case of timber poaching.

PART III

CANOPY

Chapter 17

TRACKING TIMBER

"They said that it was like stealing the crown off the Statue of Liberty or a headstone from Gettysburg."

—Danny Garcia

The Forest Service officers in Washington went to court knowing they had the evidence needed to prove that Wilke had been poaching maple. They had developed a strong case through interviews, and they planned to call former Forest Service investigator Anne Minden to the stand to explain the high demand for figured maple (wood whose longitudinal grain exhibits distinctively attractive patterning) and identify the ways in which Wilke's case mimicked so many others in the region—the tools used, the permits fabricated, the camouflaging of stumps by debris.

But their challenge lay in proving that the wood seized from the Tumwater sawmill matched samples taken from the stumps near the Maple Fire site. To present a strong-as-possible case, they sent the seized wood to Rich Cronn, the Forest Service's geneticist and molecular biologist, who works from a lab at Oregon State University in Corvallis.

Cronn's work identifies variations in the genomes of plants such as trees, shrubs, and grass. His work for the Forest Service had generally focused on studying long-term seasonal changes to tree genetics as a way of improving forest management. But in 2015 Cronn had begun receiving curious requests from Forest Service officers: Could he help them match the DNA of confiscated wood to that of stumps they had found?

The technology existed to match wood with stumps in the forest, Cronn assured them, but it was expensive. Given the consistent interest from Forest Service agents, however, Cronn and the research station lobbied successfully for funding to launch a lab dedicated to analyzing timber DNA for law-enforcement purposes. His Corvallis lab became part of a scientific crime-fighting initiative, sharing research and technology with the US Fish & Wildlife Service Forensics Laboratory, also in Oregon.

The research team conducts DNA analysis to confirm wood's original location. To do so, they grind shavings from seized blocks into sawdust, then mix it with a solution to extract the original tree's DNA. That DNA data then forms the basis of the *single-nucleotide polymorphism* (or SNP, pronounced "snip") approach, which produces a breakdown of hundreds of genetic markers within a DNA molecule, enabling the researchers to identify individual trees. SNP has been used in forensic work to match DNA to perpetrators from human crime scenes; it has proven to be even more effective in breaking down plant markers.

Cronn studies the DNA breakdowns of various ecological ranges, and he has been working to map out how species DNA evolves across geography and climate. The goal is to establish a region-wide database that law enforcement and researchers can consult when determining where seized wood might have originated. "We didn't want to make a database for [all the Douglas fir on] Vancouver Island and then a trial comes up in the Olympic

National Forest," Cronn explains. When completed, the database will be a valuable resource for the Forest Service in identifying where poached wood (or evidence such as sawdust and wood chips) originally grew, even if a stump was never found. Ideally, future researchers will be able to use the database to localize the forest origin of any given sample to within five miles, making it easier to ascertain if wood was poached from public lands.

The database has been expanding slowly but surely. Members of Cronn's team have developed collections for ranges of Douglas fir, Spanish cedar, and oak, and they are now working on other high-demand woods such as western red cedar and black walnut. But the fieldwork is tedious and time-consuming. To accelerate the database completion, the lab has partnered with an organization called Adventure Scientists, which recruits networks of scientists to collect tree samples from the backwoods of Alaska, British Columbia, Washington, and Oregon. Cronn then analyzes each sample and adds its unique profile to the database.

Already versed in gathering scientific evidence, the Adventure Scientist groups (dubbed the "Timber Tracking" team) are furnished with the tools needed to sample wood from tree boles throughout the Pacific Northwest. In the woods they extract small cores from trunks; they also gather leaves, needles, or cones from the forest floor, as well as branches from the sampled trees.

The goal is not to have one or two DNA samples from each species in the region, but rather thousands of samples from trees of the same species across their entire biome. This will create an array of DNA examples that can show where a tree may have grown, determined by the DNA of the trees around it. "A DNA sequence isn't representative of a zip code or a specific GPS point," says Cronn. "But we hope to give people precision to the level of a kilometer or 10 kilometers.

"If we have laws and they're not enforceable," Cronn continues, "they're just a suggestion. My hope is that the mill owners and the people who are buying the raw wood start asking, *Do I want to go to jail? Do I want to be tied up with anything like this?*"

In the spring of 2021, Cronn sent a note to Forest Service officers nationwide, asking about timber poaching in their areas. He received 170 responses, detailing everything from firewood theft to timber poaching for fence posts. An officer in Oregon described the meticulous behavior of a yellow-cedar poacher: all the cones and branches had been removed and only a small scattering of sawdust left behind, "like someone had vacuumed the forest floor." In the West, Cronn expects red cedar to be increasingly poached; in Alaska it's likely to be yellow cedar and Sitka spruce. "Musical instruments are our number one focus," he says. "I would like to make it possible to investigate timber theft for all tone woods." Ideally, the technology will advance to the point where traces of sawdust swept from a truck's floorboards can be tested.

In the East, the Timber Tracking team is now sampling the black walnut's range across 32 states, from Connecticut through east Texas. Because the eastern black walnut grows near rivers and streams, its range is sprawling. "It's more than up-and-coming," says Cronn of black-walnut poaching. "It's arrived. It's the new glamour wood."

In the Wilke trial, the DNA samples from the Tumwater maple-wood blocks precisely matched those from the three maple stumps in the forest. Still, Wilke had never contested the charges of wood poaching—only his responsibility for starting the forest fire.

After he testified in court at Tacoma, Cronn got back in his vehicle and listened to the rest of the trial via Bluetooth speaker as he drove home to Oregon. Cronn heard Wilke's defense lawyer invoke his name: Cronn's tree-DNA evidence was strong, she conceded, but the same could not be said for the case that Wilke had been the one to strike the match that set the woods aflame.

Cronn had pulled to the shoulder of the highway to listen. "She said, 'What would Dr. Cronn think about this?'" he remembers. "And I'm thinking, *How can I text in what I think about this?!* It's just crazy."

The Wilke case set a precedent for using tree DNA in poaching cases. Assessing the newly available detection method from his RNSP office, chief ranger Stephen Troy speculated that if all the wood seized in the Derek Hughes case had been turned, they would have simply sent some of it to Cronn's lab for analysis. Dendritic DNA profiling may not deter other poachers, admits Cronn, but he hopes it sends a warning to manufacturers and mills: "I think that Fender thinks they have vetted their mill owners really well, knowing they ask for permits. But in [Tumwater] they had one—it was just fake. I think they could easily have made five guitars from that wood."

Chapter 18

"IT WAS A VISION QUEST"

"Every single one of those shop owners would
have taken them off my hands, no problem."
—Danny Garcia

While Cronn's lab works on the database for poached timber in the United States and Canada, the illegal lumber trade in North America is dwarfed by the amount of illegal logging that occurs elsewhere. The United States holds only eight percent of the world's forests, constituting the fourth-largest global timber reserve, after Russia, Canada, and Brazil.

Most poached wood makes it into our homes from countries such as Brazil, Peru, Indonesia, Taiwan, and Madagascar, arriving as products forged from rosewood, ebony, *Dalbergia nigra*, balsa, and agarwood. Organizations like the World Bank and Interpol have estimated that the global scale of illegal logging generates somewhere between $51 billion and $157 billion annually. Thirty percent of the world's wood trade is illegal, and an estimated 80 percent of all Amazonian wood harvested today is poached. (In Cambodia that number jumps to 90 percent.)

Whatever its source, illegal wood is often sold to manufacturers in China, who then ship it to retailers and homes around the world in the form of furniture, paper products (including food packaging and napkins), construction materials, and musical instruments. Investigations have tracked down poached wood as flooring sold at Home Depot and chairs sold at IKEA.

In some cases, poached wood is just a cog in the machine of funding large-scale crime networks. The terrorist network Al-Shabab is known to traffic wood, and charcoal made from poached trees in Somalia, incorporating timber and wood products in its funding stream. Research shows this charcoal being shipped to the Gulf States, where it's used in shisha pipes. Other studies have found that charcoal used for barbecues in Belgium originated from African baobab trees. In Australia, organized-crime "firewood rings" haul in one million dollars' worth of poached Tasmanian timber each year. In Myanmar, timber poaching from Alaungdaw Kathapa National Park financed a military junta.

Even where laws exist to govern logging operations, they are often ignored or unmonitored. In most cases, experts say, trading in illegal wood is easy to perpetrate, with an almost insurmountable confluence of factors conspiring to make it difficult to stop. These include infrastructure projects that offer easy access to deep stands of forest; lack of political will to halt vast deforestation; forged documentation; and unflagging consumer demand for the cheap wood products that come from these regions. "If you deal drugs or kill an elephant, you are constantly at risk" of being caught, explains Christian Nellemann, formerly a senior officer with the United Nations Environment Programme. "If you deal timber, no one really cares."

Botanist/writer Diana Beresford-Kroeger agrees: "Trees and the forest had more legal protections 2,000 years ago in the Celtic world than today."

Some of the work being done to protect endangered tree species takes place in a nondescript, one-story building nestled in a valley at the foothills of the Siskiyou and Cascade mountain ranges. Here, the US Fish & Wildlife Service Forensics Lab works to solve environmental crimes occurring via vast, continent-spanning supply chains.

The lab, located in Ashland, Oregon, opened in 1986, primarily thanks to the efforts of a man named Terry Grosz. A Fish & Wildlife Service officer for three decades, Grosz was legendary in the world of wildlife law enforcement—the man who once floated down Humboldt County's Eel River pretending to be a salmon.

On a late-summer night in 1974, aiming to catch poachers illegally fishing the Eel River by moonlight, Grosz donned a skintight black wet suit. He was over six feet tall, with a soft, clean-shaven face that recalled a suburban uncle more than a rugged outdoorsman. *Pesce-mimicry* was a bold move, Grosz told me half a century later from his kitchen table in Colorado, but "in those days we had salmon like you couldn't believe in northern California. If you weren't careful when you drank a glass of water, you'd find a salmon in it."

It wasn't difficult to poach fish from the Eel River using large gillnets, or even by shooting them with rifles in the shallows, but the law required that anglers stop 30 minutes before sunset. Poachers routinely operated along the riverbanks far past that deadline, using glowing lures to hook salmon weighing over 60 pounds. The practice meant the salmon never reached their spawning habitats, so their stocks would eventually dwindle. But stopping the poachers challenged game wardens, even as they posted lookouts along the major roads—foreshadowing the tactics of timber poachers in the Redwoods four decades on.

After driving his truck to a spot called Singley Hole—a small break in the river where fish pooled after migrating upstream from the Pacific Ocean—Grosz donned his wet suit, tucked a citation book in its pocket, lay flat on his back on the rocks, and let the water carry him off. It was pitch dark, and he could hear the river thundering around him and see the stars through a shadowed canopy of tree crowns above. "I was trying to be quiet," he remembered. "Then I could see those lures flying in the air." Grosz snagged a lure and hooked it to his suit, then let himself be quietly reeled in to shore. There he sprang from the water and surprised the poachers with an arrest warrant.

Thus did Grosz, who would go on to spend much of his career in redwood country, make a name for himself as a wildlife officer willing to go to extremes for the cause of conservation. He moved quickly up the ranks of the Fish & Wildlife Service, eventually taking a post with its law-enforcement branch.

Grosz never considered himself much of an outdoorsman, he confides. But he was more than willing to navigate the margins of human experience to stop crime. "The position that I was in as a fish-and-game officer was not a job," he said. "It was a vision quest."

————

The Fish & Wildlife Service's fight against the wild-species trade dates to 1900, when the Lacey Act—which bans the trafficking of certain wildlife as a measure to protect food stocks—was established. Soon after, it became apparent that legislation alone would not entirely stem the tide of environmental crime. Indeed, such interdiction became increasingly complex as the world's trade networks globalized. By the 1950s, massive numbers of animals were being illegally hunted, and by the early 1960s, about 85 percent of the world's crocodiles, for instance, had been slain for the leather-goods market.

In 1973 the United States played host to a large international gathering in Washington, DC, which led to passage of the Convention on International Trade in Endangered Species of Wild Flora and Fauna (or CITES, for the tongue-twisted). CITES—and the Endangered Species Act, passed the same year—would change the lives of staffers at the Fish & Wildlife Service as well as the Forest Service, to say nothing of people living in forests around the world.

With their jobs now globalized, US Fish & Wildlife Service wardens saw their work turned upside down. No longer simply searching for poached deer or fish hidden in local truck beds, the Service was involved in prosecuting large-scale international trade deals that were flowing across American borders. Grosz became their point of contact, eventually taking on the role of endangered species officer at USFWS headquarters in Virginia in 1976.

"To be honest with you, I hated Washington," Grosz told me. But his career had embarked on what he calls a "peregrination"— a long, meandering journey to the point where law enforcement and the environment could meet. "It got more and more tangled as it went along," he recalled. "Importers were finding more and more ways and schemes to bring illegal things into the country." Grosz scrambled to catch up with criminal minds and tactics that were constantly mutating.

In 2008, the Lacey Act was expanded to include illegally harvested plants and timber. Whereas wildlife officers were accustomed to prosecuting crimes that occurred in local forests—and were adept at identifying local flora and fauna—they were not familiar with the intricacies of distinguishing among, say, various orchid species, nor with identifying which species of tree had been processed into wood panels and roof shingles stacked deep in a shipping container. The bodies of these plants and animals represented, from a forensic standpoint, one of the largest challenges the Service had ever encountered. Suddenly, broad-scale scientific investigations needed to encompass the minutiae of international trade, biological knowledge, and criminal investigations. Warehouses overflowed with seized loot, and no one had a clue what to do with it.

In the meantime, Grosz hired a former crime-scene investigator named Ken Goddard to develop ways of training officers in the field. Goddard first envisioned a potential wildlife-forensics center as simply a "bunnies and guppies lab," and he was convinced that Grosz had hired him primarily for his writing skills: if Goddard could pen a series of thriller novels set in the woods—and he had—he would be a natural at drafting crime-scene investigation manuals for inexperienced agents. But his responsibilities would soon extend far beyond manual writing.

Around the same time that Goddard was hired, a Fish & Wildlife Service agent named Tom Reilly was giving a talk about the falcon-hunting trade in Portland. Crates full of falcons had been seized at airports in Saudi Arabia, Reilly noted, but it was still very difficult to charge anyone in America with exporting the birds—there was no direct proof of where they had come from.

Listening attentively in the audience as Reilly spoke was Ralph Wehinger, whose family had lived in Ashland for so many years that it had earned a sort of ancestral town legacy. Wehinger was the 37th chiropractor in his family, and he had learned the trade like a logger or a fisherman learned his—from his forefathers cracking backs before him. He also worked tirelessly as a kind of amateur wildlife advocate, eventually setting up sanctuaries and a bird-rehabilitation center.

After Reilly's presentation, Wehinger raised his hand: "Why

don't you guys just use DNA to determine if the birds you're finding have been poached?"

"We don't have a lab yet," Reilly replied.

Wehinger resolved to change that. He knew of the perfect valley, just outside Ashland. He lobbied his state senators to approve funding for a wildlife-forensics lab, and he worked with the head of the National Audubon Society to secure the money. Eventually the $10 million required to build the lab was greenlighted by Congress, buried in an unrelated spending bill that passed quickly.

Despite knowing the extent of the international wildlife trade, Goddard had initially assumed the lab would investigate cases of poached whitetail deer, alone in their quiet valley in the woods. But that turned out not to be the case. Goddard hired Ed Espinoza, a pioneer in wood chemistry, to handle assessing the timber imports being seized by border agents and sent to the lab. "When we first started looking at [tree poaching], we were stunned," says Goddard. "We were starting to hear stories from agents in other countries about entire forests being clear-cut, and ships filled with raw trees in containerized cargo. At that point we couldn't make an identification if it was milled into planks, so we had to come up with something on that."

———

Today the forensics lab houses an expansive warehouse that Goddard and his team are working to fill with "standards"—that is, an example of each plant and animal currently being traded on the illegal market—against which they can compare incoming seizures. The Fish & Wildlife Service Forensics Lab is the world's only facility that can identify species listed as threatened by CITES, which is continually adding to its appendices of endangered species. (CITES now considers timber a crucial illegal market, on a par with elephant and rhino poaching.)

The poached wood that arrives at the forensics lab comes primarily from Africa, South America, Asia, and eastern Europe. Madagascan rosewood, for instance, is the most trafficked tree in the world. Called the "tree that bleeds," a rosewood is distinguished by its deep crimson heartwood, which makes unique faces for guitars and stringed instruments. The species has become a primary focus of CITES—and therefore of the Oregon lab. In 2012, Gibson Guitar was fined $300,000 for purchasing poached rosewood and ebony to produce fingerboards. Samples of the rosewood that had become part of those iconic instruments eventually made their way to Ashland, where they were analyzed by Ed Espinoza and his team of researchers. Similarly, agarwood often arrives at the lab in the form of wood chips or incense sticks. Its dark, aromatic resin provides a musky, earthy smell common in manufactured scents, and has become so sought after that a kilo commands up to $100,000.

The lab's labyrinthine rooms are now filled with examples of highly polished items such as guitars, rare violin pegs, and watch faces. Most of the wood is transported to the lab by government agencies like US Customs and Border Protection, the Forest Service, and the Fish & Wildlife Service. And although wood-certification entities such as the Forest Stewardship Council require chain-of-custody paperwork, that documentation is often ignored or plagiarized. Mahogany, cypress, teak, beech: all are poached, manufactured into home items, and shipped to North America—where, if luck prevails, the Fish & Wildlife Service Forensics Lab blocks it from entering the market.

The daunting quest to capture a laboratory standard for every at-risk tree species follows in the footsteps of the Swedish taxonomist Carl Linnaeus, who set out to order the world in the late

18th century. Linnaeus, who has been referred to as a poet in a scientist's body, wanted *reasons* for the beauty of nature. He saw the connections among species as a form of art, and he labored over intricate, colorful botanical drawings that were traded back and forth among polymaths in England and northern Europe. He wanted to show how they were linked, in this world of ours; to connect their beauty to ours somehow.

Linnaeus found that connection through order. He formalized what we now call the binomial classification system, used by the scientific world to determine relations and hierarchies between species and subspecies of plants and animals. Linnaeus created an ambitious if unpronounceable Latin naming system, listed in his extensive volumes of *Systema Naturae*. The *Systema* is a strikingly beautiful printed collection, with ornate cursive labels and handcrafted illustrations that showcase the biological structures of thousands of plants. In the *Systema,* the redwood burl that Danny Garcia poached is classified as *Sequoia sempervirens*. You can then track its place in the plant hierarchy back to the kingdom of Plantae, the phylum of Coniferophyta, the order of Pinales, and the genus of *Sequoia*.

Linnaeus enlisted young adventurers and traders with an interest in natural history to collect rare plant specimens from around the world and send them back to his office. Theirs was not a particularly ethical practice: his collectors often stole plants from local communities, then stuffed them in crates and smuggled them by boat to Europe. It was assumed that these plants needed to be ordered within a European framework of knowledge, that their ecological and cultural context—their purpose on Earth itself—mattered less than their function within the larger system.

Those *Systema* volumes now serve as tools in the battle to protect our environment. The Fish & Wildlife Service lab aims to irrefutably prove where snakeskins and crates of turtle meat and the delicate, colorful feathers of exotic birds have originated. It is also leading the way in applying that certainty to a new collection: a chemical database of every endangered tree in the world.

Espinoza has developed a groundbreaking method for identifying tree genus. "Up until [recently]," his boss Goddard explains, "as far as anyone in the world could go was family [the classification level above genus and species]. It's a mind-boggling discovery. Ed came up with this method for looking at the oils in wood using a DART instrument...and in doing so he almost killed himself."

Espinoza and his team use a mass-spectrometry technique to identify chemical compounds. They do this by converting oils found in the bark and wood into a gas, then injecting it into a device roughly the size of an office photocopier, known as the DART (for Direct Analysis in Real Time).

Wielding a pair of tweezers, a technician holds a tiny piece of wood—a wood chip or a bark shaving—up to a connection point on the machine, where two silver cones meet at their narrowest tips. Pinioned between those tips, the wooden assay sample is heated to 450 degrees Celsius: you can see the edges of the wood smolder and give off steam.

The vapor is then absorbed into the machine, where its molecules are analyzed. Finally, the DART sends the resulting chemical-composition data to a linked computer, where it is processed and mapped along a vector that captures a unique pattern for every species of tree, something like a fingerprint.

Once, Espinoza was running a piece of rosewood through the DART when he suddenly felt light-headed and began to experience tunnel vision. He dropped the wood and staggered away. Rosewood contains a natural insecticide, and it turned out that some of the gas had been leaking from the machine. "It was basically shutting down his brain," says Goddard.

On the day I visit the lab, I spot a chess set on a table; its pieces are about to be shaved and fed into the machine, as will parts of the box that held them. On a wall nearby hangs a photo of a trendy-looking wooden watch, marketed until recently on Instagram. The timepieces had been intercepted at the border and determined to be made from illegal timber.

Espinoza has presented his work to international wildlife-trade experts around the world. From rangers to customs officials to conservation specialists, the reaction has been consistent: this technology is a game changer. The lab's work is now conducted in tandem with some of the largest botany collections in the world. By feeding enough minuscule wooden shards through the DART, Espinoza and his team of three researchers hope to create standard vectors for every CITES-listed tree in the world—some 900 at last count. In essence, they have adapted the traditional wood library into a digital format.

Many of the wood samples that make their way into the stock-piles of the Fish & Wildlife Service Lab have been sourced from *xylaria*—wood libraries that were once maintained at the world's largest botanical gardens and in archival collections. Xylaria are generally rare now, left untouched to gather dust in the back of storage rooms. But they have proven especially useful in criminal cases related to timber theft, serving as a foundation of the anti-poaching database now being developed in Oregon.

One fall day, forensic researcher Cady Lancaster ushered me into a side room of the sprawling lab compound in Ashland. Filing cabinets lined the walls. When Lancaster slid one open, the drawer was stuffed with folders containing small slips of folded white paper, each enclosing a sliver of wood.

A few years earlier, Lancaster—then a Forest Service employee focused on the global wood-poaching trade—had been tasked with traveling the world, venturing into musty backrooms or shiny protected archives to shave splinters from chunky wood samples, many of them collected hundreds of years before. Lancaster's assignment had taken her all over the world, introducing her to archivists and historians of science who helped her locate tropical and European tree collections to flesh out the Corvallis collection. "A lot of the reference samples we have from the original blocks say *from World Fair 1903*," she said. "It's just really cool."

Cady Lancaster has unearthed book-sized slabs of wood from the Smithsonian Institution's backrooms in Washington. She has carried home slim white envelopes filled with wood slivers from the Royal Botanic Gardens, Kew, in Richmond upon Thames, England. Now those samples and countless others are filed away in Oregon, where they are steadily yielding their secrets to the DART.

Chapter 19

FROM PERU TO HOUSTON

"People are coming in to get land."

—Ruhiler Aguirre

In 2015, a shipment of poached timber sat idled in the cargo
ship *Yacu Kallpa* near the Peruvian city of Iquitos, along a bend
in the Amazon River. The ship was set to descend the Amazon
until the river opened into the Atlantic Ocean, then head north
to Tampico, Mexico, and eventually dock in Houston. Filling the
freighter's hold was wood from the Amazon, destined for factories
in the United States; there it would become floors, siding, doors.

The goal of the crew on ships like the *Yacu Kallpa* is to
complete the journey on time and without complications. In this
instance, they hoped to draw as little attention to themselves as
possible: the ship's cargo had been poached from the province
of Loreto, Peru, and the paperwork meant to authenticate the
lumber and its origin trees had been fabricated.

As the ship steamed east toward the ocean, officials in
charge of checking the paperwork against the forest home it had
identified were deep in the woods with a GPS monitor, checking

coordinates. By the time they realized the wood was incorrectly identified—and therefore likely poached—the freighter was well on its way toward American shores.

According to an investigation by the Environmental Investigation Agency (an investigatory offshoot from Greenpeace), timber merchants stocking the *Yacu Kallpa* had logged wood originally rooted in national parks and Indigenous concessions. EIA investigators are known for being meticulous, thorough, and successful. Given the vast scale of international trade, intercepting illegal wood must occur at the customs level, before the contraband has a chance to reach the market. This can be difficult for officers attempting to identify species of wood inside large shipping containers. It's not uncommon for poached wood to be hidden deep among pockets of legally sourced wood, and it's unlikely that any customs officer has the luxury to examine hundreds of planks in a single cargo ship. Officers therefore rely on good intelligence to let them know what to look out for, from where, and on what shipment. EIA provides that intelligence, and the wood often ends up in Ed Espinoza's lab.

The EIA has investigated poaching cases at the highest levels of organized crime. They once proved that a set of new—and supposedly terrorist-resistant—mahogany doors, destined for installation inside the United States Capitol, had been requisitioned from a company in Honduras suspected of poaching the mahogany from a UNESCO World Heritage Site. The order for the doors was promptly canceled.

The EIA has been at the helm of some of the most important international timber-related cases in recent years. In 2013, undercover EIA sleuths proved that the US-based company Lumber Liquidators knew that it was purchasing hardwood flooring made of wood poached from the Russian habitat of the endangered Siberian tiger. EIA teams work around the world, from South America to the forests of Romania, alongside Interpol and international law enforcement to expose the transnational timber trade. The group's research shows that 540,000 tons of rosewood, for instance, have been illegally traded since 2012—the equivalent of six million trees. In a 2015 study of imported timber from Peru, 90 percent was determined to have been illegally logged.

In the case of the *Yacu Kallpa,* the EIA sent an investigator hiking deep into the Peruvian rain forest to check the logging site claimed on the freighter's paperwork. After an arduous five days of travel, the investigator arrived at the spot to find it still completely forested. The wood aboard the ship, due for delivery to Global Plywood & Lumber in Las Vegas, had come from someplace else—somewhere it should not have.

Though the *Yacu Kallpa* switched the flag identifying its country of registration numerous times before landfall in the United States, a network of border agents had been alerted that its hold was probably stacked with illegally felled and transported wood. When the *Yacu Kallpa* docked in Houston, border agents stood ready to pounce.

They knew they would need a way to confirm that the seized wood was, in fact, poached. For that they turned to Ed Espinoza and his team in Oregon.

From the sky, the Amazon is a shag carpet of green. Only the top of the canopy is visible until you dip down onto the tarmac, where you're assailed by a wall of heat. On the outskirts of the southeastern Peruvian city of Puerto Maldonado, the economic hub of the Madre de Dios region, many small mills operate: stacks of logs and wood line the road, and the drive from the airport into the city is loud, hot, and punctuated by the sounds of chain saws and small plumes of smoke.

A bustling timber trade along the border between Peru and Brazil flows through this city. Madre de Dios borders both

Bolivia and Brazil, and the Interoceanic Highway cuts a long artery through the Amazon before entering Peru here. Vast deforestation enabled the highway's construction; along it today flow people and goods from some of the richest land in the world.

About an hour's drive southeast of Puerto Maldonado, the town of Infierno welcomes visitors with rainbow bunting stretched across a gravel road. Infierno is so named not for the sweltering heat of the surrounding rain forest but because of a pandemic flu that swept the community through missionary contact. "They felt a hot fever, so they jumped into the river," explains Ruhiler Aguirre, Infierno's community president, inside the town council meeting room. But many died of shock when they hit the cold water—so many, in fact, that missionaries called the town Infierno after their floating bodies: "The missionaries said, *Wow, this looks like hell.*"

Peru's land is managed by a concession system. As with the Forest Service, the National Park Service, and other conservation authorities in the United States, each type of Peruvian concession— ecotourism, forestry, conservation—pursues a different objective. Many conservation and ecotourism concessions have been returned to Indigenous people as their traditional territory to manage. Infierno controls a conservation concession of just under 15,000 acres of traditional Ese'eja territory down the Rio Tambopata, which the town partly funds thanks to an annual stipend of 60,000 Peruvian soles ($15,000) from the federal government for social assistance, education, retirement, and health fees.

To visit the conservation concession that April, a group consisting of myself, journalist Milton Lopez Tarabochia, Aguirre, and three guides pilot a fiberglass boat down the Rio Tambopata on a foggy, gray morning. After an hour or so, the river journey morphs into a hike in the woods, which take us past a forest guardian's hut on our way to another boat, one that would ferry us across a lake. The river-shore hut had been built to house a revolving crew of guardians. Featuring a grass roof and a broad patio stacked with deck chairs and propane canisters, the hut (and others like it in the surrounding area) is a form of forest protection: Infierno's conservation land had become a target for timber poachers. From 2017 to 2018, some of the tallest trees in the concession were poached—felled and processed on site into plywood and planks, which were then shipped to city markets.

During our slow paddle across Tres Chimbadas Lake, we see crocodiles and otters and the brightest birds imaginable. As we slow, turning toward shore and a rough brush of bright green foliage, a copper glint shines through the leaves. As the boat's pilot coasts to dock, the copper's source becomes apparent: large stacks of processed wood waits along the shore. They had been placed there for easy access, to be transported to markets in Puerto Maldonado. "Even with all their equipment," Aguirre says, pointing toward the lumber, "these people can't hide the wood. This is the beginning."

We move past the stack of wood, stepping on it like terraced stairs off the boat, and walk six miles into the jungle, the narrow path covered in thick roots and leaves that had fallen from the trees towering above us. The poaching along the path is hard to miss: the trail itself had been carved out of the forest—not by Infierno or its forest wardens, but by poachers using it to shuttle wood to transportation points.

The concession land is carpeted with tropical cedars, mahogany, and a hardwood known as estoraque. But the most targeted trees are ironwoods, which have become essential in the production of parquet floors. A foot-long chunk of ironwood sells for three soles, while an entire tree fetches about 3,000 ($775). This wood is often exported to Asia for manufacturing before finding its way to Europe and North America. But standing in their home forests in Peru, ironwoods are rich in flora and fauna, their branches furnishing homes for macaws and the harpy eagle, one of the biggest raptors in the Amazon. Every poaching site we visited had zeroed in on shi-huahuaco, a type of ironwood and the tallest tree in its region of the Amazon. In 2018, experts predicted that the rate of shihuahuaco poaching would devastate the population of those trees within a decade. That year alone, 141,000 cubic feet of shihuahuaco were illegally harvested. The concession had been home to 52 towering shihuahuaco in 2015, but only 41 were still standing by 2018.

Passing by a standing shihuahuaco, we crane our heads as far back as possible but still cannot see its crown. Most shi-huahuaco specimens reach their peak height at 1,000 years old. The tree is prized for its tough wood, not unlike mahogany, and it provides traditional medicine and food to the Ese'eja peoples in Infierno. It is also an anchor species in a landscape that Ese'eja ancestors once moved through, a resting spot between destina-tions. The shihuahuaco has slowly become one of the favorite targets of South American poachers. "One, two, three, four iron-wood down," Aguirre counts, as we traveled the jungle path.

The first poaching site we cross was veiled by a scrim of branches. Dried brown fronds had been placed over the stump of what was once a towering shihuahuaco. In front of the stump a clearing had been scoured out, creating a staging area where the tree could be parceled up for transport down the road in a small motorized cart.

Deeper into the concession we find another clearing, this one even larger and stacked with milled rectangular blocks of wood blanketed by sawdust. The wood is stacked in such a way that it resembles the steps leading down to an amphitheater stage, showcasing the place where the tree stood, where it was cut into blocks, where it was shaped and sanded.

The strict government guidelines around what can and cannot be done on designated parcels of conservation-concession land mean that small communities such as Infierno are tasked with monitoring impossibly large areas of lush terrain. They shoulder the blame if a tree goes missing from their land, a giant slipping between their fingers. "Basically, if we lose one foot it becomes a problem—*my* problem—because the wood is in our hands," Aguirre says, standing atop a timber stack. "We take care of this wood, so the process keeps going.

"People are coming to our land to set up different activities," he continues. "Right now there is a family established [on the concession]. They took a piece of land and are doing timber extraction. In English, those are called squatters."

Squatting is a problem throughout the Amazon. With so much economic migration occurring throughout Peru, Brazil, and Bolivia, many people simply settle down wherever they wind up and begin their lives anew, without asking landowner permission. One edge of Infierno's conservation concession borders a vein of the Interoceanic Highway, and it has been difficult to prevent small settlements from springing up within the stands of trees there. The highway has sliced through some of the region's pristine biodiversity, making room for large trucks to transport resources to ports. At the same time, it has opened up a world of trade and an opportunity for people to travel for employment.

Puerto Maldonado is a hub for immigrants—the city welcomes some 300 new residents each day. This, paired with an influx of migration from Peru's Andes region into the country's south for work, means that regions like Madre de Dios are now swamped with people striving to make a living, looking for anything that might pay the bills, and seeking out a place to live. Every morning on the streets of Puerto Maldonado, men line up hoping to be picked for daily logging jobs. Often they are hired and sent somewhere to fell trees illegally.

"The way to get money is the illegal way," says Aguirre. "It happens forever. It's a cultural thing, where people think they can move onto land and take wood because it's something they need."

He estimates that the 11 poached shihuahuaco (totaling about 30,000 board feet) were worth around 10,000 soles ($2,600).

Infierno's guardians have at times removed the structures and belongings of squatters on their concession, but it never keeps them away for long. They are frustrated and angered by the poaching, but by migration as well—too many people from the northern regions of Peru were coming in for work, they said. But it's unlikely those loggers know their work is illegal.

From across the lake, wardens can sometimes hear the poaching happening on the concession land, and they listen as chain saws and machinery buzz to life in the dark. Once, someone from Infierno approached the family doing the poaching but grew worried for their safety.

Theirs is a situation very similar to the challenges rangers face in the redwoods. "If you don't find [them] in the place where they're doing the activity, [you] can't do anything," says Aguirre. "So we had to find them doing the logging." But this is difficult—wardens are a boat ride away from the concession land, and they have no cellphone reception in the forest. In order to pin the poachers to the wood, and the site, they would need to devise an elaborate scheme.

—————

In March of 2018, Aguirre and the Infierno town council began securing the paperwork and partnerships required to take action the next time a forest warden heard a chain saw start up in the woods. "We worked as a community, as a team," Aguirre explains. They arranged for Puerto Maldonado's police to be on standby well in advance: only six "environmental police" officers work across the entire region of Madre de Dios, and there are thousands of acres of lush jungle to patrol. Nor is timber poaching the sole environmental crime plaguing the department: the mass deforestation of the Amazon has simultaneously opened doors for illegal mining, oil and gas drilling, and the cultivation of drug crops such as *coca*.

The community also asked a television journalist to travel with police to the concession whenever they were called in. By including broadcast news, they hoped the public would be outraged by the images they saw on TV, squelching the market for poached wood at local mills.

The guardians then installed sound-sensitive alarms in trees and received training in how to use GPS devices that would pinpoint poaching locations. The alarms were concealed just above "poacher height" in the foliage of various shihuahuaco trees throughout the district. Whenever a chain saw sprang to life, the alarm tipped off the guardians, who alerted Aguirre back in Infierno; he would then notify the environmental police and the TV journalist, hopefully summoning them to the tree that had "gone off" in time to film—and stop—the destruction.

When the sound of a chain saw ripped through the night in the spring of 2018, the team was ready. The wardens texted Aguirre, who boarded the boat and floated downstream to the wardens' outpost, where he was joined by the police and the journalist. The group then glided silently across the lake, landing softly on the shore and walking unobtrusively through the woods to reach the poachers.

This particular band of poachers was an industrious lot: they had already taken the wood to a makeshift mill erected in the middle of the forest, where the police caught them and charged them with illegal logging. The case would become a high-profile drama in Puerto Maldonado's court system.

When I visited the concession land with Aguirre only two months after the bust, we made our final stop at the sawmill. It was still standing, the forest clearing littered with discarded saws and chains. Remnants of the loggers' presence persisted: a ripped T-shirt left behind on a table, a box of soap, an empty tin can.

Aguirre worries that squatters are still selling poached wood to mills in Puerto Maldonado. The processed timber abandoned in the forest, meanwhile, has been allowed to decompose back into the earth—part of the Ese'eja's mandate is to keep the forest as intact as possible. The rangers visit often to make sure the wood remains.

The squatters' encampment on Infierno's land set my head spinning: it looked just like the remnants of the tent encampment I had seen while shadowing NRO Luke Clarke in Nanaimo, British Columbia. As in Canada, so in Peru: both encampments were home to those who could not afford any other option, despised and necessary at once. Locals had been left at the mercy of powerful economic tides in both forests, working in logging because they had the connections or because they couldn't do anything else, taking single trees off protected land in order to get by.

Our day on the poacher's trail ended in the clouds. Aguirre and I climbed the steep staircase to the Posada Amazonas lodge located deeper in the woods, an oasis in an intricate jungle. We sat in the lodge's immaculate restaurant, drinking fresh fruit juice as a breeze blew paper-thin curtains around us. The resort, owned by the Ese'eja from Infierno, runs on solar power and boasts a deck built into the sky, a canopy-topping lookout from which you can survey the vast carpet of green waving in the breeze. This is what Aguirre wants Ese'eja land to be used for: he wants to spotlight its glory, to welcome people and show them some of the brightest birds in the world flitting along the riverbanks and perching on the eroded shores, to point out where the snakes and spiders are and laugh at their startled reactions.

In the end, says Aguirre, Infierno is faring well. "We are doing conservation, we are doing tourism. We hold our culture in those things."

Chapter 20

"WE TRUST IN THE TREES"

"Our forest is our pharmacy. It's our market."

—Jose Jumanga

The most active timber region in Peru is in the department of Ucayali. Its river-port city of Pucallpa ships some of the world's largest trees to manufacturers around the world. The city's riverfront along the Rio Ucayali plays host to a bustling market that sells fruit, live animals, and homewares, and it's lined by ships and boats that spew dark exhaust into the air. Beyond the city boundaries, small settlements and villages follow the Ucayali, a transportation link that connects the Amazon to the outside world.

The Indigenous communities surrounding Pucallpa number about 300, dotting the shores of the Ucayali and stretching inland deep into the Amazon basin. But Indigenous peoples manage only 25 percent of the land in Ucayali; the government of Peru has handed over the rest of the forest to private logging companies. The forest communities in Ucayali experience deforestation from their concessions in much the same way that

Infierno does. Often, sections of the Amazon there are cleared of timber for agricultural use by poachers who log and burn the land, far away from community hubs.

Raul Vasquez, a researcher with the Upper Amazon Conservancy, has spent time with Amazonian forest communities, working as a timber monitor on Indigenous lands. Many logging companies have lobbied for him to lose his job—his life, and the lives of his family members, have been threatened because of his work. "One day someone got out of a car and threatened my wife," Vasquez tells me during a meeting in his office. "So it's complicated."

Vasquez is well-versed in the deforestation that happens around the region. In the field he often stumbles across trunks and pieces of wood that have been prepared for sale to an exporter or manufacturer. He says that local communities need more investment in order to manage their own land and protect it from corporate greed. Much of that investment, he predicts, would go to simple technology solutions such as drones and GPS devices.

While the packed water taxis leaving from Pucallpa transport dry goods and people from the city, there is also a form of subsistence that comes from the rain forest itself: "We trust in the trees," says Jose Jumanga, president of Comunidad El Naranjal. "It's not the plant that protects us, it's the life inside the plant. This is the essence of the forest, and it takes care of us. For that reason, we protect the forest. For the life that is inside."

El Naranjal first tracked timber poaching in the 1990s, a ripple effect of increased migration to the region. After an economic crisis in the northern regions of Peru, immigration to Ucayali spiked, in part because the land was prime for cultivation. Outside El Naranjal, that led to logging and clearing of land for ranching and farming. The cultivation included not just food and palm-oil production but *coca* plants—a small runway had been built to shuttle *coca* illegally from the depths of El Naranjal's forest.

In 2000, Comunidad El Naranjal gained access to a GPS system. (About 10 percent of the Indigenous communities surrounding Pucallpa have been given GPS to monitor the landscape they maintain.) El Naranjal intended to digitally map the boundaries of their community, but the GPS system also revealed where land had been logged without consent. Altogether, El Naranjal's wardens estimate, about 9,000 acres of forest land have been cleared without their authorization since the 1990s. One group of people, linked to a religious movement, wantonly established an entire community in the woods; around it they posted signs welcoming visitors to El Naranjal land the sect is occupying illegally.

Only two wardens from El Naranjal patrol the conservation concession, and they do so on foot. They are often vulnerable, coming face-to-face with not only drug traffickers but also game poachers, who take deer and small pigs from the forest to sell in illegal markets. On a trip outside El Naranjal, I watched as the lush woodlands—the forest thickets dense with shihuahuaco (and other ironwoods), caoba, and ishpingo trees—gave way to tracts of open, rolling green land. "Our forest is our pharmacy, it's our market," says Jumanga. "It's a place [from which] we take our wood—but for [building] houses, not for sale. All the resources are decreasing with deforestation."

The madera trees that once grew in El Naranjal's precincts often end up as exteriors for houses. Mahogany trees have been felled to make dyes. Formerly forested land is now used to graze cattle. From his vantage point in the village, Jumanga frequently hears chain saws or spots plumes of smoke rising from the forest.

If the trees aren't burned once the land is cleared, poached logs are transported to Pucallpa, where they are given falsified paperwork and exported. Jumanga and timber investigators like Vasquez estimate that only about 40 percent of the trees poached from their land are taken to Pucallpa now, a significant decrease from the early 1990s, when the number was closer to 90 percent. Still, the depredations haven't been stopped, nor have the challenges to doing so changed: there is ever the unlikelihood of catching poachers in the act, the fear of what it might mean to approach someone and ask them to stop, the relentless market demand that invariably makes the payoff warrant the risk.

In early 2018, the nonprofit activist group Regional Organization of AIDESEP in Ucayali (ORAU) granted Comunidad El Naranjal a surveillance drone, which allows them to track logging activity in corners of their 1.5-million-acre concession so remote that they are inaccessible to forest wardens. Overflying the tree canopy—which at points drops away to reveal bare, brown patches of land—the drone has opened a portal showing the community which parts of their land are most vulnerable to timber poaching.

Despite the hard evidence furnished by the drone's aerial footage, poaching reports made from El Naranjal have elicited scant government response. Though the air around us swayed with a dulcet breeze, Jose Jumanga's tone grew increasingly urgent throughout our afternoons together: "We are worried about this," he understates.

Chapter 21

CARBON SINKS

"Forest means life."   —Jose Jumanga

The trees that grow in the forests outside El Naranjal and in the concessions of Infierno, alongside the old-growth contained in the parks of North America, are standing during several international crises: an environmental one, in which we are on track to lose 20 percent of the world's species in our lifetimes; and a social and economic one, in which marginalized peoples who live in exploited environments are left scrambling to eke out a living. Timber poaching exists at the confluence of these challenges.

Forests are one of our greatest protectors against climate change. Their continued destruction hastens global warming, the loss of biodiversity, and the extinction of species. Forests pull about one-third of the world's human-caused carbon—about 1.5 times more than the US produces—from the atmosphere each year. But big trees (those with trunks more than 21 inches in diameter) have a particular power to curb climate change: because they have reached full root, bark, and canopy development, they store more carbon than younger, still-developing new growth. In 2018, a study found that the world's widest trees hold about half of all the carbon stored in global forests. Thus logging old-growth forests deprives the planet of the "carbon sink" provided by old trees catching carbon and storing it indefinitely. In a double-whammy effect on the environment, the resulting carbon imbalance is only exacerbated by the carbon released through industry.

Old-growth redwood, cedar, and Douglas fir are stalwart carbon sinks: the trees rooted along the west coast of Vancouver Island, for example, hold some of the highest amounts of stored carbon in the world. Not only does the tree canopy play a crucial role in carbon storage, so too does the forest floor around the tree. Studies from the Forest Service show that the decomposition of trees and branches is a necessary element of recycling nutrients and storing carbon in a forest. Old-growth trees are resilient, too: they withstand fire better, and their dense, moist ecosystems tamp down the heat that dries out stands of trees elsewhere.

But it would be foolish to suggest that the trees in North America or Europe are as critical to our climate as those nearest the equator. Tropical trees constitute one of the world's primary defenses against climate change. The Amazon is home to more than 16,000 tree species—16 times the number in North America—and new species are continually being identified in the region. These trees absorb 120 million metric tons of carbon each year, provide essential habitat for endangered wildlife, and contribute to a food-production system that sustains millions of people.

We can find in stories from these forested regions an echo of the actions that took place centuries ago in North America. Corporations continue to cut as much as they can in the Amazon today—by some accounts, the area of a soccer field every minute. Just as they have in the Pacific Northwest, local economies have become tethered to the opportunities provided by over-harvesting trees. Entire communities have been destabilized by the promises and perils of lumber money, including Indigenous groups that have been systematically threatened, displaced, and sometimes killed in order to gain free access to their land. But in newspaper reports, Brazilian president Jair Bolsonaro (who dubs himself "Captain Chain Saw") rebuffs the protests of environmental organizations: "This land is ours, not yours." His predecessor, Luiz Inácio Lula da Silva, voiced a particularly anti-colonial sentiment: "I don't want any gringo asking us to let an Amazon resident die of hunger under a tree."

By the summer of 2021, so much damage had been done to the Amazon rain forest that it began to emit more carbon than it stores. Rain forests grow on expansive peatlands; when these are disturbed to prepare the ground for agriculture, carbon dioxide accumulated over millennia is released into the atmosphere. Indonesia's peatlands presently release more CO_2 than the state of California. In South America, environmental organizations have conducted satellite research substantiating their worst fear: that logging has now penetrated the "untouched core" of the Amazon basin, reaching Indigenous conservation areas deep inside Brazil.

In theory, rising temperatures could lead trees to metabolize faster, thus absorbing carbon more quickly. But scientists have discovered that warming temperatures can also cause them to respirate—that is, release—carbon faster than they are able to incorporate it in photosynthesis. If temperatures rise too high, a tree can be damaged at the molecular level. Climate change has also created a tinder-dry forest floor, and increasingly warm forest environments. This causes a forest fire (sparked by humans or lightning) to spread farther and faster than in the past, making it harder to control or extinguish. The net effect is an existential threat for old-growth trees: for coast redwoods to survive, those burls may be forced to reproduce more often than they otherwise would have.

Climate change also stunts the growth of trees like the Douglas fir, which shut down in extreme temperatures and stop sequestering carbon. "If the trees aren't putting away the carbon, we have to think, *How do we readjust?*" asked Forest Service geneticist and molecular biologist Rich Cronn after a heat wave passed through the West in the summer of 2021. "It's forcing us to readjust."

In response, forest experts are now openly debating a sort of botanical Hobson's choice: Which tree species, and which ecosystems, are worth saving?

The beeps that alerted Infierno to the flutter of a chain saw on its concession land are heard across South America and Asia now, adopted throughout rain forest communities as a method to protect forests against poaching.

In recent years, other new technologies—often funded by non-governmental organizations such as the Rainforest Alliance—have also been developed. (These systems are sometimes even paid for by timber and palm-oil companies, which want to maintain accurate sourcing information about their suppliers.) Improved radar techniques, for example, have made it easier to capture images of the rain forest canopy through dense cloud. This, paired with advances in satellite imaging, means more consistent monitoring of the rain forest, with more alerts and warnings being sent. Higher-resolution images make it possible to focus on individual trees, handing some control back to smaller communities that have iconic, towering old-growth in their forests.

At the Fish & Wildlife Forensics Lab in Ashland, I sit with Cady Lancaster in front of what looks like a large scanner. The lab has begun to dabble in identification through fluorescence. In broad daylight, the bark and wood of many species are almost indistinguishable from that of others; under fluorescence, by contrast, wood becomes a vibrant display of glowing neon chromosomes. Every species has a unique fluorescent pattern that emerges under a black light, like a fingerprint. When examined by a microscope this fluorescence shines, etched into the ridges of bark and tree rings, flowing in circles and patterns throughout the trunk. In some cases there is an ombré effect, indicating a shift in the wood's structure.

Lancaster slides a cross section of black locust wood under her black light: the outer edges are still covered in bark, and parts of the sample have been damaged by an ax cut. Instantly the natural fluorescence of the wood springs to life, bright yellows and greens stitched together along the rings and ridges. As Lancaster zooms in, new colors emerge from the wood's pores—tiny dots and lines surfacing on a ligneous canvas. Indeed, fluorescent identification *is* something of an art form—an indication of the surprising hidden beauty all around us. (Attesting to that fact, Lancaster has made postcards of her favorite fluorescent patterns.)

When I visited the lab in Ashland in autumn 2019, Ed Espinoza too had been contemplating the intersection of beauty and trees. Having watched a YouTube video of a man "playing" a slice of redwood on a record player (allowing the groove of each tree ring to emerge as a warped soundscape), Espinoza had become preoccupied with communicating the voices of trees. But the filmed technique didn't seem right to him: Does a tree's voice truly come from its rings?

Espinoza has since gotten a bit closer to finally hearing the music in wood. In 2019 David Bithell, a professor of digital arts at Southern Oregon University, approached him with an intriguing proposition: if Espinoza would share the data collected from a poached tree, Bithell's students could feed it into a computer and generate electronic music.

On a crisp fall evening, Espinoza and Lancaster visited Professor Bithell's music class, which had been mining the data from a number of species files submitted by the forensics lab. As interpreted by the SOU students' software, the data produced a combination of low, thrumming noises, repetitive and hypnotic, each tree emitting its own vibration frequency. Espinoza leaned back in a folding chair and interlaced his fingers behind his head, listening as his trees came to life.

For all the advances in technology, rangers remain the primary method of deterring timber poaching. That system is a North American conservation legacy, which has been exported and established around the world. As poaching has become a lucrative global trade, a military-style ranger system has sprung up to combat it: poaching is now countered by increasingly armed rangers, cultivated local-informant networks, and heavily patrolled conservation areas. With intense pressure to protect some of the world's most endangered species, this "fortress" model of conservation has been adopted reflexively in reserves in Africa and Asia, from armed guards protecting Siamese rosewood in Thailand to park rangers monitoring some of the last remaining rhino and elephants.

When Irish conservationist Rory Young wrote his iconic "Field Manual for Anti-Poaching Activities" in 2014—one of the world's only handbooks for rangers on the ground—he declared poaching a complex crime that "must be understood within a cultural context." Whether it's the hacking of redwoods for their burls or the slaughter of elephants for their tusks, Young implored his readers to try to understand the communities from which poachers spring as a means of stopping them. Prevention, he wrote, includes "dealing with socio-economic factors that encourage poaching."

Notably, Young used the ancient woods of England to make his point: the Sheriff of Nottingham's fatal error, he wrote, was allowing Robin Hood to become ingrained in the community: "[Hood's] greatest asset was the support of the people." Despite the power of the sheriff and his bailiffs, Robin Hood and his gang of merry men always won: "This is the classic mistake made over and over again," Young noted. "It does not matter how many heavily armed and well-equipped men with battle skills are sent in, if they can't find the poacher it is just a waste of time, effort and money."

Young avoided heralding the advent of new technology to solve poaching. Men tracking poachers on foot across a forest floor are likely to be just as efficient as remote-sensing apparatus, he wrote, and no "wonder weapon" will ever put an end to poaching. Instead, wrote Rory Young, the thing most likely to dissuade poaching is deep knowledge of the area itself: "Know thine enemy!"

Back in Orick, California, the ranger-as-police-officer approach has germinated a good deal of resentment. Given the makeup of the western world's conservation system—large swaths of nature left unpopulated, humans separated from nature—the woods lack local guardians who might discourage a poacher from poaching. Instead, there are police. Steve Frick, the former Peanut Convoy driver, says he thinks many of the town's problems could be eased if the park got "rid of men who pull guns."

The Outlaws speak openly and fervently about the injustice with which they believe the Park Service reigns over Orick. How they feel monitored, judged, attacked. "They harass people, pull people over all the time," Cherish Guffie says, standing in Terry Cook's front yard.

"They're just trying to pin shit on me," Cook adds.

Much of this can be written off as an expected sentiment from the accused. Some of it, however, is rooted in fair criticism—it's common for rangers to start investigations by focusing on those with prior-arrest records, whether they have reason to or not. But that anger can quickly metastasize: "They don't want me to be their enemy," Cook says. "They won't have any trees left...I got a saw over there that'll cut down any trees they got."

Chapter 22

IN LIMBO

"But I will also add this: I retired years ago."

—Derek Hughes

The fractured friendships of The Outlaws are ever in flux—it is a small circle, and one in which suspicion and paranoia reign. Loyalty remains the utmost goal of any relationship. This emotional atmosphere is stoked by the park's practice of seeking out informants in exchange for dropping charges on smaller infractions. Friendships and alignments tend to change quickly; insults are freely exchanged. One day I was told that Chris Guffie was staying with Terry Cook. The next time I visited, Guffie was no longer welcome in Cook's house.

The last time I spoke with Guffie on the phone, he was pacing the floor, getting worked up. He had recently spent time working on oil rigs in Wyoming, but when we spoke he was homeless. "Of course all of us would have been working still, we work for our money, we're not out for a handout or anything," he says that afternoon, from a house in the coastal California town of Trinidad. "But when we go out to try and make a few bucks for the family, we get penalized—thrown in jail or whatnot." Guffie sees himself as a scapegoat for any type of crime that happens in the park. He suspects that a lot of people offer to provide information on him when they get pulled over by park rangers. "If you give up Chris Guffie," he says, "they let you walk free." Since that exchange, my connections with Guffie have been sporadic—I have not heard from him since the fall of 2020. After he missed a court date for a case related to poaching wood, a warrant for his arrest was issued that July.

Chris's father, John Guffie, eventually sold his house in Orick and moved south to McKinleyville. Throughout Chris's troubles with the park, John has supported his son. They share certain characteristics: anger toward the Park Service, frustration at lack of opportunity. John labels Orick a town "just for druggies" now. And all the better, he says, echoing a conspiracy that many in Orick have peddled to me: eventually the town will empty out, enabling the Park Service to take it over fully.

Despite Derek Hughes's previous relationship with Danny Garcia, the ties that bound the two men seem to have frayed. "I never did what he did, with live trees and all that...that's just bullshit if you ask me," Hughes says, insisting that he stuck to dead-and-down wood. Cook and Guffie have both distanced themselves from Garcia's case: "He took it to extremes, he cut the fuck outta that tree," Cook said in his yard.

"Yeah, that was not okay," Cherish added. "We were mad at him for that."

Garcia has worked hard to turn his life around. In this he has been aided by local rancher Ron Barlow, who helped Garcia fulfill his required community-service hours and take the first few tentative steps back into the conventional workforce. "Sometimes you see something in somebody and you think, *We have to fix this*," Barlow says. Garcia now works at sawmills in and around Eureka, where he rents a two-bedroom house and lives with his girlfriend and their daughter.

"She's more for the environment," Garcia says of his partner. "And I would be too, I think. But over the years I got this shitty side that I acquired because of hatred toward the park." He no longer visits Orick, he says, and he has not been allowed to enter the park since his May 2014 conviction. He still gets angry when speaking about the case, though, believing that Larry Morrow received a softer sentence than he did and that the park rangers were out to get him. As we talk, Garcia shares something his girlfriend told him: "She says if I had been born one generation earlier, it would have all been acceptable."

Derek Hughes waited three years for the outcome of his trial—the court date was pushed back time and time again from 2018 to 2021. During that time he sat in limbo, taking on small landscaping and general laborer jobs and taking care of his mother, Lynne Netz. "[The park] fired my mom and I find it very, very difficult every day to cope with that because my mom loved her job," he says. "And I'm the reason she doesn't have it." He hopes to leave Orick one day soon.

As the Hughes case plodded on, RNSP ranger Branden Pero placed hidden cameras pointing toward the wood on Hidden Beach. On the beach, it was easy to spot logs that had been sliced into and cleaved, chunks of wood poached. But the beach was also piled high with logs and decaying driftwood that were sun-bleached white and black, appearing like small rolling hills of wood. In the images captured by the cameras, Pero saw locals visiting with trucks at night, cutting firewood and transporting it out. He couldn't make out who they were.

While Hughes's case sat in limbo, Pero moved on to a job with the US Forest Service, though Humboldt remained his operating base. "I don't know what he thinks is going to happen," Pero said of Hughes before departing for his new assignment.

On an August day in 2021, Derek Hughes and his lawyer stood before a judge inside Humboldt County Superior Court. Hughes had maintained his innocence until the month previous, claiming there were plenty of people who matched his height and might have been driving a truck like the one captured in the surveillance video taken at May Creek in 2018. He was certain that the Park Service's case against him was flimsy: "All they have," he said, "is a blurry-ass photo of an outline of a body that looks kinda like mine." Having grown disillusioned with his first court-appointed attorney, he had cycled through a total of four public defenders as the case was continually delayed.

But Hughes had a change of heart in July 2021, when he struck a plea deal with prosecutors. Most of the charges against him were dropped and he pleaded guilty only to felony vandalism. He had hoped the judge would be amenable to downgrading that conviction to a misdemeanor; Hughes wanted to keep possession of his four guns, of which a felony conviction would strip him.

In a statement given for his probation hearing in August, Hughes suggested he'd been "set up" by Larry Netz, his step-father, who had recently moved out of Lynne's home. During the sentencing trial, however, the judge made it clear that he was unimpressed by the allegation: "I don't see any remorse being reflected for this event, and really no acknowledgement of wrong-doing," he told the court. He had hoped that Hughes would have reached out to the RNSP to offer restitution, demonstrating that he recognized the ramifications of the crime beyond its mere monetary consequences. "I don't see that," said the judge. "I see a concern with not being able to pack a gun."

The Park Service prosecutor had lobbied for the maximum sentence, including a $10,000 fine and a full ban from park property. But Hughes is considered unlikely to become a repeat offender, and both his mother and his sister had recently been diagnosed with cancer. Hughes argued that he didn't deserve an absolute ban from the park, especially because he needed to travel the highway through it to ferry his family to their medical appointments. In the end he was sentenced to two years' probation, 400 community-service hours, a fine of $1,200, and banishment from the park other than traversing it.

Derek Hughes stood and faced the judge as he prepared to receive this sentence. On the wall to his right hung the official state crest of California, carved in redwood.

AFTERWORD

In British Columbia, echoes of the War in the Woods still reverberated 30 years later as I was finishing this book. In July and August of 2021, environmental activists gathered in a region of Vancouver Island called Fairy Creek, where the provincial government had granted logging company Teal Jones a parcel of old-growth timber. Deep in the rainy woods, protesters lodged themselves in branches, erected platforms in the canopy, unfurled banners from tree limbs, and lay down in front of logging machinery. The protests soon eclipsed the records set at Clayoquot Sound in 1993: more than 1,000 people were arrested at the site over a five-month period.

I watched news from the Fairy Creek Blockade avidly from my home in the province's interior. I live in logging country, where the 2019 closure of the sawmill near my town triggered some 200 layoffs in the region. The community council is now tasked with diversifying an economy that was, until recently, stable and booming.

Just outside my front door, down a narrow road that leads onto an even-narrower dirt road, a small forest of Douglas fir, hemlock, and western red cedar stands along the banks of the North Thompson River. The forest is managed by the Wells Gray Community Forest Corporation, which established it in 2004 during a previous decline in the region's logging. The community forest now surrounds the town of Clearwater and is managed for various uses—including logging. Profits from the wood harvested from the forests that surround my house usually go back to local charities and groups. The enterprise is a driver of local employment, and also local culture.

Community forests like the Wells Gray have been slowly spreading across British Columbia, a small testament toward sustainable use in a stand of trees. In fact, the province's first community forests were developed in 1998 in the wake of the War in the Woods. In a report released in 2021, the British Columbia Community Forest Association found that community forests create more than twice as many jobs as those run by independent industry, and that half of the community forests in British Columbia are managed by Indigenous communities or in partnership with them.

Fifty-nine community forests now dot the province. Most of them are run by municipalities of fewer than 3,000 people, who manage the timber and the products from it themselves, leasing the land from the provincial government on a long-term basis. More than 1,500 people in the province earned some income from community forests in 2021—not only in logging but in firefighting, trail building, and scientific study.

Community forests provide one solution for how forest-management practices might better represent forest communities themselves. At the very least, such forests help stave off the sort of resentment that erupted in Orick recently when the Save the Redwoods League signed an agreement with Redwood National and State Parks that allows limited logging as part of a park-restoration venture called the Rising Redwoods project. Watching heavy equipment clearing brush on park land—as well as chain saws felling and trimming some trees—confused and enraged many residents of Orick.

"And they call *us* the bad guys for taking the dead stuff," Derek Hughes says. "I thought it was all about saving the redwoods.

"They make the rules but don't follow them," he adds. "So what does that tell everybody else?"

The custom of managing one's own community forest extends well beyond British Columbia, of course. Locals staffed the concessions I visited in Peru, and some community forests in Mexico are so successful that experts have proposed them as a global model. Notably, community forests have reduced poverty in thousands of forest regions.

As we witnessed with the Sunshine Coast Community Forest, poaching still happens in community forests. But the factors contributing to it might be lessened if poachers knew they were stealing from neighbors, not from an anonymous park administration. Paired with community guardians like those in the forests in Infierno, it might be possible to make the stakes for poaching too high. For some, it could even strengthen the bonds of community—as Hughes says, if he knows the person pulling him over, life in town might be less tense, more amicable.

This requires a new approach to conservation, and would ask rangers to put down their guns. Globally, forest-governance experts have begun advocating for conservation policies that give communities control over the forest that surrounds them— even if that means harvesting trees. And those who propose conservation projects without taking human use into account receive sharp pushback. In 2020, a team of 100 economists and scientists released a report imploring governments to conserve 30 percent of the world's land and water by 2030. But theirs was the fortress model of conservation, devoid of humans, and their plan implied that tourism would fill the gap of resource extraction. In response, conservation researchers and social scientists around the world volunteered their critiques: "This paper reads to us like a proposal for a new model of colonialism," thundered one appraisal.

Separating nature from human use has never kept it safe—the Charter of the Forest reckoned with this knowledge, addressing issues that remain relevant hundreds of years later. Instead, the result is a legacy of generational trauma: "I got a husband who never recovered from it," says Nadine Bailey, who read a statement at President Clinton's Portland summit in 1994. "He tried and did a number of things, but he was never himself and it took a part of his soul. The broken promises to the rural community, that's what makes people lose hope."

Ultimately, protecting trees is a question of belonging: Where are you from? What do you understand of these woods? "And if you want to get to the bare bones," says Derek Hughes, "all this land belongs to the Yurok."

ACKNOWLEDGMENTS

This book is hinged to the generosity of people who opened up their lives to me, particularly in Orick, California. Notably, the book would not exist without the deep honesty of Danny Garcia, Derek Hughes, Lynne Netz, Terry Cook, Cherish Guffie, Chris Guffie, and John Guffie, who repeatedly answered very personal questions about their lives, and who demonstrated commitment, kindness, and transparency as I tried to tell a complete and nuanced story.

Chief Ranger Stephen Troy, Branden Pero, Laura Denny, Special Agent Steve Yu, and the rest of the team at Redwood National and State Parks walked me through their work in the park and answered many follow-up questions with grace and patience. The National Park Service and Department of the Interior were remarkably swift in providing me with the documents I needed to track the stories in this book. Andy Coriel and Phil Huff, with the US Forest Service in Oregon, were also invaluable sources. In British Columbia, Luke Clarke and Denise Blid allowed me a seat in their patrol truck and were patient with my questions. I was granted on-the-ground access to the US Fish & Wildlife Service Forensics Laboratory in Oregon, and I thank Ed Espinoza, Ken Goddard, and Cady Lancaster for guiding me through their research processes. Thanks as well to Rich Cronn, for sharing his expertise on the amazing world of tree DNA.

Mackenzie Brady Watson believed in me and this story from

the beginning, even during the times when I was unsure of the path before me. I thank her and Aemilia Phillips, at Stuart Krichevsky Literary Agency, for their constant support, advice, and guidance. Daisy Parente at Lutyens & Rubinstein advocated for my work in the UK, and I'm eternally grateful that my book will find its way onto shelves there.

Tracy Behar and Ian Straus at Little, Brown Spark, and Jennifer Croll at Greystone Books, gave me thoughtful, challenging, and smart edits, and helped this book find its way through the woods when it was lost. Thank you as well to Huw Armstrong at Hodder & Stoughton, for sharing this book with UK readers and seeing its connection to England's forests. Thank you to Allan Fallow for astute copyedits that improved the manuscript greatly. My gratitude to the marketing teams at Little, Brown Spark, Greystone Books, and Hodder & Stoughton for their work in helping this book find an audience.

Thank you to Jeffrey Ward for his genius map. It's an honor to include your work alongside my writing. Thanks as well to Jen Monnier for her fastidious fact-checking. Cassidy Martin was an excellent research assistant and transcriptionist and now knows as much about tree poaching as I do.

The COVID-19 pandemic prevented me from traveling as extensively as I had planned throughout the Pacific Northwest. I'm incredibly grateful for the time I was able to spend on the streets of Humboldt County and the many phone calls and messages its residents accepted from me throughout my time writing this book. The Humboldt County Historical Society gave me access to its archives and a beautiful research space, as well as many interesting conversations and insights. I was educated on Humboldt's kindness, open spirit, and deeply forgiving nature by those in community drug response throughout the region. Teresa Janowski at Humboldt Superior Court helped me navigate the system there and was invaluable in my hunt for transcripts and documents. Every source in this book has been a candid, cooperative, welcoming force in my life, and this is, I hope, a slice of their reality.

I was welcomed in many communities throughout the Madre de Dios and Ucayali regions of Peru during my fieldwork there in 2018. Thank you to Milton Lopez Tarabochia for ensuring my reporting went smoothly, and for being a smart and funny companion on riverboats and jungle hikes. Thank you to Rosa Baca with the Federación Nativa del Rio Madre de Dios y Afluentes, Ruhiler Aguirre and Comunidad Nativa de Infierno, Comunidad Nativa Belgica, and Comunidad El Naranjal for sharing your knowledge and experiences with me, and for allowing me to pitch a tent on your land. Thank you as well to Julia Urrunaga with the Environmental Investigation Agency and Raul Vasquez of the Upper Amazon Conservancy, who bravely research Amazonian timber theft at great risk.

Much of my work has been influenced by the dedicated scholarship of academics who study the history of conservation in North America. While this extensive body of work is noted in the bibliography, I want to send particular thanks to Dr. Karl Jacoby, Dr. Erik Loomis, and Dr. Dorceta Taylor, whose writing gave me solace, motivation, inspiration, and confidence in my own observations. Thank you to Virginia, for her diaries. I am also incredibly indebted to Amelia Fry's oral-history archives, held at the Bancroft Library at the University of California Berkeley. Thank you to the librarians at the Thompson Rivers University library, who are masters of the interlibrary loan.

The portions of this book that sprang from my magazine and online journalism have been influenced by wonderful editors, including Michelle Nijhuis, Rachel Gross, and Brian Howard. The research and reporting that constitute this book received financial assistance from the National Geographic Society, the Society of Environmental Journalists' Fund for Environmental Journalism, the Canada Council for the Arts, and the Alberta Foundation for the Arts.

The writing in this book was enriched through workshops including the Banff Centre's Mountain and Wilderness Writing Workshop and the Bread Loaf Environmental Writers' Conference. Many thanks to my mentors—John Elder, Marni Jackson, and Tony Whittome—and fellow writers at these conferences, which boosted my confidence and inspired me in many ways. As well, thank you to Jessica J. Lee and Sarah Stewart Johnson, who generously provided me with templates and encouragement as I wrote the proposal for this book.

I am lucky to have thoughtful, smart, and creative friends and mentors, many of whom provided editorial notes and ethical insight throughout the writing process. There are not enough thanks in the world for Allison Devereaux, Margaret Herriman, Jamie Hinrichs, Michelle Kay, Stephen Kimber, Karen Pinchin, and Sandi Rankaduwa.

My family—Danielle Bourgon and Gareth Simpson, Daryl Bourgon and Trina Roberge, Lisa and Archie Huizing, and my grandparents Rick and Charlyne Bourgon—have always supported me and my work unequivocally, and I thank them for their patience and enthusiasm when this book took me to strange places. My work springs from the opportunities they have given me, and thrives with the unwavering support I receive from Simon Corkum.

I wrote *Tree Thieves* with the following words from James Agee's *Let Us Now Praise Famous Men* as a guiding light:

> To come devotedly into the depths of a subject, your respect for it increasing in every step and your whole heart weakening apart with shame upon yourself in your dealing with it: To know at length better and better and at length into the bottom of your soul your unworthiness of it: Let me hope in any case that it is something to have begun to learn.

The entire process has been an honor.

GLOSSARY

ALASKAN MILL: a small-scale, portable sawmill that incorporates a chain saw; it can be used by one or two operators to mill logs into lumber.

BIRD'S EYE: a distinctive and highly valuable pattern disrupting the smooth lines of a wood's grain.

BLANK: a block of wood that can be turned (that is, carved) into an artisanal product.

BOARD FEET: a unit of measurement for lumber volume. One board foot equals the volume of a board that is one foot long, one foot wide, and one inch thick.

BUCK: to cut a tree into lengths after it has been felled.

CAT SKINNER: the operator of a Caterpillar tractor.

CHECKING WOOD: cutting a small piece off the trunk of a tree with an ax or chain saw in order to inspect the grain inside.

CHOKER: a short length of wire cable wrapped around one or more logs so that they can be hauled to a landing and loaded.

CLEAR-CUT: to log a forest area until every tree it once contained has been cut down and removed.

COPPICE: an area of woodland where trees are cut back to ground level to stimulate growth, as well as to provide firewood and timber.

CORD: 128 cubic feet of wood.

CUTBLOCK: a specific area, with strictly defined boundaries, that has been authorized for harvesting.

CUTLINE: the result of logging done so as to make a straight clearing through a forest, often as a way of delineating a property boundary.

DEAD-AND-DOWN: a tree that has died and fallen of natural causes.

DESIRE LINE: an unofficial trail tramped down in a forest by repeated foot traffic.

DUFF: the vegetative matter—such as leaves, twigs, dead logs, and so on—that is often found covering the ground in a forest.

EDGER: a device whose saws are used to straighten and smooth rough lumber.

FELL: the act or process of cutting down a tree.

FELLING WEDGE: a thick piece of plastic that helps keep a standing tree from pinching a chain-saw bar, typically causing the tree to fall in the direction of the notch cut.

FIGURE: the appearance of a wood's grain.

GOOSEPEN: a hollowed-out tree trunk that is large enough to accommodate a grown man.

GREEN CHAIN: a lumber-delivery system utilized inside a sawmill. "Pulling green chain" refers to collecting the final product of the mill and moving it at a controlled rate to be graded and sorted.

HEARTWOOD: the dead central wood of a tree (sometimes also called "duramen").

MILL: a factory where logs are processed into lumber.

MUSIC WOOD: sometimes referred to as "tonewood," the raw material used to create the top, back, sides, fretboard, and bridge of a stringed instrument.

NOTCH: a cut made in a tree or downed log in the shape of the letter *V*.

SALVAGE SITE: a parcel of land, containing dead or dying timber, that is sold for logging.

SECOND-GROWTH: woodland growth that replaces harvested virgin forest.

SETTING CHOKERS: attaching log-hauling cables to logs for transport.

SHAKES AND SHINGLES: terms used interchangeably for split, rectangular pieces of tapered wood. A *shake* can also describe a crack and separation of wood between the rings.

SLAB: an outer piece taken from a log or timber.

SNAG: a dead-standing or dying tree.

SPLIT LUMBER: wood that has been split along its grain rather than being bucked into rounds.

STAG CAMP: a temporary logging worksite, usually built alongside a river, with bunkhouses and a cookhouse.

STAND: a community of trees, all relatively similar in size, age, and distribution.

TIMBER CRUISER: a person who specializes in surveying a stand of timber and estimating how many board feet of marketable lumber it contains.

TREE-SPIKING: the act of driving large nails into the trunk of a tree with the goal of destroying either the logging equipment that attempts to harvest it, the quality of the wood thus obtained, or both.

TURNING: the action of shaping wood with a lathe.

NOTES

CHAPTER 1: CLEARANCES

10 $1 billion: This figure was determined by an Associated Press report in 2003 and is the most current estimate. It is also widely used in literature surrounding the illegal-timber trade.

10 $100 million: This figure comes from a Forest Service study conducted in the 1990s. A more recent study has not been conducted, and this is the number that Forest Service officials continue to reference.

CHAPTER 2: THE POACHER AND THE GAMEKEEPER

15 eleven people: Records show that the vast majority of poachers were (and generally still are) men, but this particular day included women.

15 Forest of Corse: Otherwise known as Corse Lawn.

18 "verderers": Three swanimotes still exist in England today, with verderers courts: Forest of Dean, Epping Forest, and New Forest.

19 wearing a woman's dress: He had dubbed himself "Lady Skimmington."

CHAPTER 3: INTO THE HEART OF THE COUNTRY

23 "starve but for the game": From Karl Jacoby's *Crimes Against Nature*.

24 frontier tradition: In what would become a great act of irony, the conservationist John Muir's father even poached wood from the land surrounding their family homestead in Wisconsin.

28 efficient machinery: Despite improvements in the safety of logging, forestry is still one of the most dangerous professions in the world. In social scientist Louise Fortmann's interviews with logging families, one wife told her that she thanked God every day that her husband hadn't yet been killed. In 1976, deaths in the forest industry exceeded those of police and fire workers in the same area.

29 eugenicists: A monument to Madison Grant previously located at Prairie Creek State Park in Humboldt was removed in June 2021.

CHAPTER 4: A LUNAR LANDSCAPE

37 tour of northern California by the US House of Representatives Subcommittee on National Parks and Recreation: Led by Chairman Wayne Aspinall of Colorado, who was presented with a gavel made of redwood.

CHAPTER 5: REGION AT WAR

46 Retraining programs had been delayed: One trimworker explained to the summit attendees: "At 14, I learned our heritage."

52 "tree huggers": Notably, the roots of this term are in eco-feminist protests that were part of India's Chipko movement in the 1970s.

CHAPTER 6: THE GATEWAY TO THE REDWOODS

66 ligno-tubers: There are some instances of saplings being poached as well, but there is no paperwork documenting that charges were ever brought against anyone for doing this.

CHAPTER 7: TREE TROUBLES

71 In 1970, Cook's family: There is some uncertainty as to the exact year when the Cooks arrived in Orick. Terry Cook and Garcia independently remembered the date as 1970, but in 2019 Cook told me he had lived in Orick for 58 years, which would mean the family had arrived in 1961. There is no record of the Cooks in Orick before 1970.

73 which John shut down: John Guffie shut the company down because he didn't want to keep up with the heavy costs of continual machinery-and-technology upgrades.

74 "until he worked with me": Garcia disagrees and says he learned more, generally, from "being around people."

74 standing tree: After speaking with Guffie twice, I was unable to contact him again.

76 returned to Orick: Both still have warrants out for their arrest in Washington.

CHAPTER 9: THE TREES OF MYSTERY

88 "50 years ago or longer": Local historians have confirmed that taking burls has been happening for at least this long.

90 woman's wig and sunglasses: Guffie neither confirmed nor denied this was him.

93 Klamath burl shop named Trees of Mystery: The owners of the Trees of Mystery did not respond to interview requests about the case.

CHAPTER 10: TURNING

100 get rid of the campground altogether: I have not found confirmation that this meeting happened, but it has become local lore.

CHAPTER 11: BAD JOBS

103 poverty rate was 26 percent: Humboldt's poverty rate is consistently higher than the California average.

CHAPTER 12: CATCHING AN OUTLAW

119 phone message from Danny Garcia: Garcia denies that this meeting and conversation happened, but it is logged in the official paperwork.

CHAPTER 13: IN THE BLOCKS

126 with Danny Garcia: Garcia says that he knows Hughes well. "But I mean, pretty much anything we did together, we did together."

126 "gave us the beach": Garcia agrees on this note.

127 "You really gonna fill his shoes?": Pero does not recall this interaction.

134 "Blew my mind!": Hughes had indeed been investigated as part of this case, as he would be in other cases in the years following.

134 was fired: Lynne Netz suggests that Stephen Troy told other department heads across both parks not to hire her for seasonal work.

CHAPTER 14: PUZZLE PIECES

138 overseas markets: Kamada remembers walking into a hotel room and finding 1,000 succulents, stored in coolers and littering the furniture.

139 set of old keys: Hughes later told investigators that he got the keys from "a guy" he did not remember.

143 Charles Voight: Voight declined to be interviewed for this book.

CHAPTER 16: THE ORIGIN TREE

159 Shawn Williams: Nicknamed "Thor," and he looks the part: he has a broad, superhero frame and long, wavy blond hair.

CHAPTER 17: TRACKING TIMBER

167 'What would Dr. Cronn think about this?': Cronn's work is notably trustworthy. The odds that his DNA analysis returned coincidental results are infinitesimal: one in one undecillion (a *1* followed by 36 zeros).

CHAPTER 18: "IT WAS A VISION QUEST"

171 from his kitchen table in Colorado: Terry Grosz passed away a year after we spoke, in 2019.

CHAPTER 21: CARBON SINKS

200 Irish conservationist Rory Young: In April 2021, Young was killed along with two journalists as they traveled with an anti-poaching brigade in Burkina Faso.

CHAPTER 22: IN LIMBO

206 "set up": There is no indication of this in the case files, but Hughes insists that Larry was involved in providing information about him to the rangers.

BIBLIOGRAPHY AND SOURCES

EPIGRAPH

Williams, Raymond. "Ideas of Nature," in *Culture and Material-ism: Selected Essays*. London: Verso, 2005.

PROLOGUE: MAY CREEK

Author's personal notes and photographs, Sept. 2019.
Court filings, "People of the State of California v. Derek Alwin Hughes." Case no. CR1803044, accessed Dec. 2020.
Goff, Andrew. "Orick man arrested for burl poaching, meth." *Lost Coast Outpost* (Eureka, CA), May 17, 2018.
Pero, Branden. Interviews with the author, Sept. 2019 and Sept. 2021.

CHAPTER 1: CLEARANCES

Alvarez, Mila. *Who owns America's forests?* U.S. Endowment for Forestry and Communities.

"Arkansas man pleads guilty to stealing timber from Mark Twain National Forest." *Joplin (MO) Globe,* Apr. 21, 2021.

Atkins, David. "A 'Tree-fecta' with the Oldest, Biggest, Tallest Trees on Public Lands." United States Department of Agriculture Blog, Feb. 21, 2017. https://www.usda.gov/media/blog/2013/08/23/tree-fecta-oldest-biggest-tallest-trees-public-lands.

Benton, Ben. "White oak poaching on increase amid rising popularity of Tennessee, Kentucky spirits." *Chattanooga (TN) Times Free Press,* Apr. 4, 2021.

Carranco, Lynwood. "Logger Lingo in the Redwood Region." *American Speech* 31, no. 2 (May 1956).

Closson, Don. Interview with the author, Sept. 2013.

Convention on International Trade in Endangered Species of Wild Flora and Fauna. *The CITES species.* https://cites.org/eng/disc/species.php.

"800-year-old cedar taken from B.C. park." Canadian Press, May 18, 2012.

Frankel, Todd C. "The brown gold that falls from pine trees in North Carolina." *Washington Post,* Mar. 31, 2021.

Friday, James B. "Farm and Forestry Production and Marketing Profile for Koa *(Acacia koa),*" in *Specialty Crops for Pacific Islands*, Craig R. Elevitch, ed. Holualoa, HI: Permanent Agriculture Resources, 2010.

Golden, Hallie. "'A problem in every national forest': Tree thieves were behind Washington wildfire." *Guardian* (London), Oct. 5, 2019.

Government of British Columbia. Forest and Range Practices Act. https://www.bclaws.gov.bc.ca/civix/document/id/complete/statreg/00_02069_01#section52.

International Bank for Reconstruction and Development/The World Bank. *Illegal Logging, Fishing, and Wildlife Trade: The Costs and How to Combat It.* Oct. 2019.

Kraker, Dan. "Spruce top thieves illegally cutting a Northwoods cash crop." *Marketplace,* Minnesota Public Radio, Dec. 23, 2020.

Neustaeter Sr., Dwayne. "The Forgotten Wedge." Stihl B-log. https://en.stihl.ca/the-forgotten-wedge.aspx.

North Carolina General Statutes. 14-79.1. *Larceny of pine needles or pine straw.* https://www.ncleg.net/EnactedLegislation /Statutes/HTML/BySection/Chapter_14/GS_14-79.1.html.

Pendleton, Michael R. "Taking the forest: The shared meaning of tree theft." *Society & Natural Resources* 11, no. 1 (1998).

Peterson, Jodi. "Northwest timber poaching increases." *High Country News* (Paonia, CO), June 8, 2018.

Ross, John. "Christmas Tree Theft." *RTE News,* aired Nov. 8, 1962. https://www.rte.ie/archives/exhibitions/922-christmas -tv-past/287748-christmas-tree-theft/.

Salter, Peter. "Old growth, quick money: Black walnut poachers active in Nebraska." Associated Press, Mar. 10, 2019.

Stueck, Wendy. "A centuries-old cedar killed for an illicit bounty amid 'a dying business.'" *Globe and Mail* (Toronto), July 3, 2012.

Sullivan, Olivia. "Bonsai burglary: Trees worth thousands stolen from Pacific Bonsai Museum in Federal Way." *Seattle Weekly,* Feb. 10, 2020.

"Three students cited in theft of rare tree in Wisconsin." Associated Press, Mar. 30, 2021.

Trick, Randy J. "Interdicting Timber Theft in a Safe Space: A Statutory Solution to the Traffic Stop Problem." *Seattle Journal of Environmental Law* 2, no. 1 (2012).

Troy, Stephen. Interview with the author, Aug. 2018.

United States Department of Agriculture. *Who Owns America's Trees, Woods, and Forests? Results from the U.S. Forest Service 2011–2013 National Woodland Owner Survey.* NRS-INF-31-15. Northern Research Station, 2015.

United States Department of Agriculture, Forest Service, Southwestern Region. "Public Comments and Forest Service Response to the DEIS, Proposed Prescott National Forest Plan." Albuquerque, NM, 1987.

Van Pelt, Robert, et al. "Emergent crowns and light-use complementarity lead to global maximum biomass and leaf area in Sequoia sempervirens forests." *Forest Ecology and Management* 375 (2016).

Wallace, Scott. "Illegal loggers wage war on Indigenous people in Brazil." nationalgeographic.com, Jan. 21, 2016.

Wilderness Committee. "Poachers take ancient red cedar from Carmanah-Walbran Provincial Park." May 17, 2012. https://www.wildernesscommittee.org/news/poachers-take-ancient-red-cedar-carmanah-walbran-provincial-park.

Woodland Trust. "How trees fight climate change." https://www.woodlandtrust.org.uk/trees-woods-and-wildlife/british-trees/how-trees-fight-climate-change/.

World Wildlife Fund. "Stopping Illegal Logging." https://www.worldwildlife.org/initiatives/stopping-illegal-logging.

CHAPTER 2: THE POACHER AND THE GAMEKEEPER

Bushaway, Bob. *By Rite: Custom, Ceremony and Community in England 1700–1880*. London: Junction Books, 1982.

Hart, Cyril. *The Verderers and Forest Laws of Dean*. Newton Abbot: David & Charles, 1971.

Hayes, Nick. *The Book of Trespass: Crossing the Lines That Divide Us*. London: Bloomsbury Publishing, 2020.

Jones, Graham. "Corse Lawn: A forest court roll of the early seventeenth century," in Flachenecker, H., et al., *Editionswissenschaftliches Kolloquium 2017: Quelleneditionen zur*

Geschichte des Deutschen Ordens und anderer geistlicher Institutionen. Nicolaus Copernicus University of Toruń, Poland, 2017.

Langton, Dr. John. "The Charter of the Forest of King Henry III." St. John's College Research Centre, University of Oxford. http://info.sjc.ox.ac.uk/forests/Carta.htm.

———. "Forest vert: The holly and the ivy." *Landscape History* 43, no. 2 (2022).

Million, Alison. "The Forest Charter and the Scribe: Remembering a History of Disafforestation and of How Magna Carta Got Its Name." *Legal Information Management* 18 (2018).

Perlin, John. *A Forest Journey: The Story of Wood and Civilization*. Woodstock, VT: The Countryman Press, 1989.

Rothwell, Harry, ed. *English Historical Documents, Vol. 3, 1189–1327*. London: Eyre & Spottiswoode, 1975.

Rowberry, Ryan. "Forest Eyre Justices in the Reign of Henry III (1216–1272)." *William & Mary Bill of Rights Journal* 25, no. 2 (2016).

Standing, Guy. *Plunder of the Commons: A Manifesto for Sharing Public Wealth*. London: Pelican/Penguin Books, 2019.

Standing, J. "Management and silviculture in the Forest of Dean." Lecture, Institute of Chartered Foresters' Symposium on Silvicultural Systems, Session 4: "Learning from the Past." University of York, England, May 19, 1990.

St. Clair, Jeffrey. "The Politics of Timber Theft." *CounterPunch* (Petrolia, CA), June 13, 2008.

Tovey, Bob, and Brian Tovey. *The Last English Poachers*. London: Simon & Schuster UK, 2015.

CHAPTER 3: INTO THE HEART OF THE COUNTRY

Akins, Damon B., and William J. Bauer, Jr. *We Are the Land: A History of Native California*. Oakland: University of California Press, 2021.

Andrews, Ralph W. *Timber: Toil and Trouble in the Big Woods*. Seattle: Superior Publishing, 1968.

Antonio, Salvina. "Orick: A Home Carved from Dense Wilderness." *Humboldt Times* (Eureka, CA), Jan. 7, 1951.

Barlow, Ron. Interview with the author, Oct. 2021.

Carlson, Linda. *Company Towns of the Pacific Northwest*. Seattle: University of Washington Press, 2003.

Clarke Historical Museum. *Images of America: Eureka and Humboldt County*. Mount Pleasant, SC: Arcadia Publishing, 2001.

Clarke Historical Museum interpretive gallery materials. Sept. 2019.

Coulter, Karen. "Reframing the Forest Movement to end forest destruction." *Earth First!* 24, no. 3 (2004).

Drushka, Ken. *Working in the Woods: A History of Logging on the West Coast*. Pender Harbour, BC: Harbour Publishing, 1992.

Fry, Amelia R. *Cruising and protecting the Redwoods of Humboldt: Oral history transcript and related material, 1961–1963*. Berkeley, CA: The Bancroft Library, Regional Oral History Office, 1963.

Fry, Amelia, and Walter H. Lund. *Timber Management in the Pacific Northwest Region, 1927–1965*. Berkeley, CA: The Bancroft Library, Regional Oral History Office, 1967.

Fry, Amelia R., and Susan Schrepfer. *Newton Bishop Drury: Park and Redwoods, 1919–1971*. Berkeley, CA: The Bancroft Library, Regional Oral History Office, 1972.

General Information Files, "Orick," HCHS, Eureka, California.

Gessner, David. "Are National Parks Really America's Best Idea?" *Outside*, Aug. 2020.

Harris, David. *The Last Stand: The War Between Wall Street and Main Street over California's Ancient Redwoods*. New York: Times Books, Random House, 1995.

Jacoby, Karl. *Crimes Against Nature: Squatters, Poachers, Thieves, and the Hidden History of American Conservation*. Berkeley: University of California Press, 2001.

Lage, Ann, and Susan Schrepfer. *Edgar Wayburn: Sierra Club Statesman, Leader of the Parks and Wilderness Movement: Gaining Protection for Alaska, the Redwoods, and Golden Gate Parklands*. Berkeley, CA: The Bancroft Library, Regional Oral History Office, 1976.

LeMonds, James. *Deadfall: Generations of Logging in the Pacific Northwest*. Missoula, MT: Mountain Press Publishing Company, 2000.

McCormick, Evelyn. *Living with the Giants: A History of the Arrival of Some of the Early North Coast Settlers*. Self-published, Rio Dell, 1984.

———— *The Tall Tree Forest: A North Coast Tree Finder*. Self-published, Rio Dell, 1987.

"Millionaire Astor Explains About His Famous Redwood." *San Francisco Call*, Jan. 15, 1899.

O'Reilly, Edward. "Redwoods and Hitler: The link between nature conservation and the eugenics movement." From the Stacks (blog). New-York Historical Society Museum and Library. Sept. 25, 2013. https://blog.nyhistory.org/redwoods-and-hitler-the-link-between-nature-conservation-and-the-eugenics-movement/.

Peattie, Donald Culross. *A Natural History of North American Trees*. San Antonio, TX: Trinity University Press, 2007.

Perlin, John. *A Forest Journey: The Story of Wood and Civilization*. Woodstock, VT: The Countryman Press, 1989.

Post, W. C. "Map of property of the Blooming-Grove Park Association, Pike Co., Pa., 1887." New York Public Library Digital Collections. https://digitalcollections.nypl.org/items /72041380-31da-0135-e747-3feddbfa9651.

Rajala, Richard A. *Clearcutting the Pacific Rain Forest: Production, Science and Regulation*. Vancouver: UBC Press, 1999.

Rutkow, Eric. *American Canopy: Trees, Forests, and the Making of a Nation*. New York: Scribner, 2012.

Sandlos, *Hunters at the Margin: Native People and Wildlife Conservation in the Northwest Territories*. Chicago: University of Chicago Press, 2007.

Schrepfer, Susan R. *The Fight to Save the Redwoods: A History of the Environmental Reform, 1917–1978*. Madison: University of Wisconsin Press, 1983.

Shirley, James Clifford. *The Redwoods of Coast and Sierra*. Berkeley: University of California Press, 1940.

Speece, Darren Frederick. *Defending Giants: The Redwood Wars and the Transformation of American Environmental Politics*. Seattle: University of Washington Press, 2017.

Spence, Mark David. *Dispossessing the Wilderness: Indian Removal and the Making of the National Parks*. Oxford: Oxford University Press, 2000.

St. Clair, Jeffrey. "The Politics of Timber Theft." *CounterPunch* (Petrolia, CA), June 13, 2008.

Taylor, Dorceta E. *The Rise of the American Conservation Movement: Power, Privilege and Environmental Protection*. Durham, NC: Duke University Press, 2016.

Tudge, Colin. *The Tree: A Natural History of What Trees Are, How They Live, and Why They Matter*. New York: Crown, 2006.

United States Department of the Interior. "The Conservation Legacy of Theodore Roosevelt." Feb. 14, 2020. https://www.doi .gov/blog/conservation-legacy-theodore-roosevelt.

Warren, Louis S. *The Hunter's Game: Poachers and Conservationists in Twentieth-Century America*. New Haven, CT: Yale University Press, 1999.

Widick, Richard. *Trouble in the Forest: California's Redwood Timber Wars*. Minneapolis: University of Minnesota Press, 2009.

CHAPTER 4: A LUNAR LANDSCAPE

Anders, Jentri. *Beyond Counterculture: The Community of Mateel*. Eureka, CA: Humboldt State University, August 2013.

Associated California Loggers. "Enough Is Enough," 1977. Humboldt State University, Library Special Collections. https://archive.org/details/carcht_000047.

Barlow, Ron. Interview with the author, Oct. 2021.

British Columbia Ministry of Forests and Range. "Glossary of Forestry Terms in British Columbia." March 2008. https://www.for.gov.bc.ca/hfd/library/documents/glossary/glossary.pdf.

Buesch, Caitlin. "The Orick Peanut: A Protest Sent to Jimmy Carter." *Senior News* (Eureka, CA), Aug. 2018.

California Department of Parks and Recreation. "Survivors Through Time." https://www.parks.ca.gov/?page_id=24728.

California State Parks. "What Is Burl?" https://www.nps.gov/redw/planyourvisit/upload/Redwood_Burl_Final-508.pdf.

Center for the Study of the Pacific Northwest. "Seeing the Forest for the Trees: Placing Washington's Forests in Historical Context." https://www.washington.edu/uwired/outreach/cspn/Website/Classroom%20Materials/Curriculum%20Packets/Evergreen%20State/Section%20II.html.

Childers, Michael. "The Stoneman Meadow Riots and Law Enforcement in Yosemite National Park." *Forest History Today*, Spring 2017.

Clarke Historical Museum. "Artifact Spotlight: Roadtrip! The Orick Peanut," July 1, 2018. http://www.clarkemuseum.org /blog/artifact-spotlight-roadtrip-the-orick-peanut.

Cook, Terry, and Cherish Guffie. Interview with the author, Sept. 2019.

Coriel, Andrew, and Phil Huff. Interview with the author, July 2020.

Curtius, Mary. "The Fall of the 'Redwood Curtain.'" *Los Angeles Times*, Dec. 28, 1996.

Daniels, Jean M. United States Department of Agriculture, Forest Service. "The Rise and Fall of the Pacific Northwest Export Market." PNW-GTR-624. Pacific Northwest Research Station, Feb. 2005.

DeForest, Christopher E. United States Department of Agriculture, Forest Service. "Watershed Restoration, Jobs-in-the-Woods, and Community Assistance: Redwood National Park and the Northwest Forest Plan." PNW-GTR-449. Pacific Northwest Research Station, 1999.

Del Tredici, Peter. "Redwood Burls: Immortality Underground." *Arnoldia* 59, no. 3 (1999).

Dietrich, William. *The Final Forest: The Battle for the Last Great Trees of the Pacific Northwest*. New York: Penguin, 1992.

Food and Agriculture Organization of the United Nations. "North American Forest Commission, Twentieth Session, State of Forestry in the United States of America." St. Andrews, New Brunswick, Canada, June 12–16, 2000. http://www.fao.org /3/x4995e/x4995e.htm.

Frick, Steve. Interview with the author, Sept. 2019.

Fry, Amelia R. *Cruising and protecting the Redwoods of Humboldt: Oral history transcript and related material, 1961–1963*. Berkeley, CA: The Bancroft Library, Regional Oral History Office, 1963.

Fryer, Alex. "Chipping Away at Tree Theft." *Christian Science Monitor,* Aug. 13, 1996.

General Information Files, "Orick," HCHS, Orick, California.

Gordon, Greg. *When Money Grew on Trees: A. B. Hammond and the Age of the Timber Baron.* Norman: University of Oklahoma Press, 2014.

Guffie, John. Interview with the author, Oct. 2020.

Harris, David. *The Last Stand: The War Between Wall Street and Main Street over California's Ancient Redwoods.* New York: Times Books, Random House, 1995.

Humboldt Planning and Building. Natural Resources & Hazards Report, "Chapter 11: Flooding." Eureka, CA, 2002.

Johnson, Dirk. "In U.S. Parks, Some Seek Retreat, but Find Crime." *New York Times,* Aug. 21, 1990.

Lage, Ann, and Susan Schrepfer. *Edgar Wayburn: Sierra Club Statesman, Leader of the Parks and Wilderness Movement: Gaining Protection for Alaska, the Redwoods, and Golden Gate Parklands.* Berkeley, CA: The Bancroft Library, Regional Oral History Office, 1976.

"Loggers Assail Redwood Park Plan." *New York Times,* Apr. 15, 1977.

Loomis, Erik. *Empire of Timber: Labor Unions and the Pacific Northwest Forests.* Cambridge: Cambridge University Press, 2015.

Nelson, Matt. Interview with the author, Mar. 2020.

Pryne, Eric. "Government's Ax May Come Down Hard on Forks Timber Spokesman Larry Mason." *Seattle Times,* May 5, 1994.

Rackham, Oliver. *Woodlands.* Toronto: HarperCollins Canada, 2012.

Rajala, Richard A. *Clearcutting the Pacific Rain Forest: Production, Science and Regulation.* Vancouver: UBC Press, 1999.

Redwood National and State Parks. "About the Trees." Feb. 28, 2015. https://www.nps.gov/redw/learn/nature/about-the -trees.htm.

Redwood National Park. "Tenth Annual Report to Congress on the Status of Implementation of the Redwood National Park Expansion Act of March 27, 1978." Crescent City, CA, 1987.

"Redwood National Park Part II: Hearings before the Subcommittee on National Parks and Recreation of the Committee on Interior and Insular Affairs, House of Representatives. H.R. 1311 and Related Bills to establish a Redwood National Park in the State of California. Hearings held Crescent City, Calif., April 16, 1968, Eureka, Calif., April 18, 1968." Serial No. 90-11. Washington, DC: US Government Printing Office, 1968.

"S. 1976. A bill to add certain lands to the Redwood National Park in the State of California, to strengthen the economic base of the affected region, and for other purposes: Hearings Before the Subcommittee on Parks and Recreation of the Committee on Energy and Natural Resources." Washington, DC: US Government Printing Office, 1978. *(Via private library of Robert Herbst, Aug. 2020.)*

Speece, Darren Frederick. *Defending Giants: The Redwood Wars and the Transformation of American Environmental Politics.* Seattle: University of Washington Press, 2017.

Spence, Mark David. Department of the Interior, National Park Service, Pacific West Region. "Watershed Park: Administrative History, Redwood National and State Parks." 2011.

Thompson, Don. "Redwoods Siphon Water from the Top and Bottom." *Los Angeles Times*, Sept. 1, 2002.

Vogt, C., E. Jimbo, J. Lin, and D. Corvillon. "Floodplain Restoration at the Old Orick Mill Site." Berkeley: University of California Berkeley: River-Lab, 2019.

Walters, Heidi. "Orick or bust." *North Coast Journal of Politics, People & Art* (Eureka, CA), May 31, 2007.

Widick, Richard. *Trouble in the Forest: California's Redwood Timber Wars.* Minneapolis: University of Minnesota Press, 2009.

CHAPTER 5: REGION AT WAR

Bailey, Nadine. Interview with the author, Sept. 2019.

Bari, Judi. *Timber Wars.* Monroe, ME: Common Courage Press, 1994.

Carroll, Matthew S. *Community and the Northwestern Logger: Continuities and Changes in the Era of the Spotted Owl.* New York: Avalon Publishing, 1995.

Dumont, Clayton W. "The Demise of Community and Ecology in the Pacific Northwest: Historical Roots of the Ancient Forest Conflict." *Sociological Perspectives* 39, no. 2 (1996): 277–300.

"Forks: Timber community revitalizes economy." Associated Press, Dec. 21, 1992.

Glionna, John M. "Community at Loggerheads Over a Book by Dr. Seuss." *Los Angeles Times,* Sept. 18, 1989.

Greber, Brian. Interview with the author, June 2020.

Guffie, Chris. Interview with the author, Sept. 2020.

Fortmann, Louise. Interview with the author, June 2020.

Harter, John-Henry. "Environmental Justice for Whom? Class, New Social Movements, and the Environment: A Case Study of Greenpeace Canada, 1971–2000." *Labour* 54, no. 3 (2004).

Hines, Sandra. "Trouble in Timber Town." *Columns,* December 1990.

Loomis, Erik. *Empire of Timber: Labor Unions and the Pacific Northwest Forests*. Cambridge: Cambridge University Press, 2015.

Loomis, Erik, and Ryan Edgington. "Lives Under the Canopy: Spotted Owls and Loggers in Western Forests." *Natural Resources Journal* 51, no. 1 (2012).

Madonia, Joseph F. "The Trauma of Unemployment and Its Consequences." *Social Casework* 64, no. 8 (1983): 482–88.

"Northwest Environmental Issues." C-SPAN, aired Apr. 2, 1993. https://www.c-span.org/video/?39332-1/northwest-environmental-issues.

O'Hara, Kevin L., et al. "Regeneration Dynamics of Coast Redwood, a Sprouting Conifer Species: A Review with Implications for Management and Restoration." *Forests* 8, no. 5 (2017).

Pendleton, Michael R. "Beyond the threshold: The criminalization of logging." *Society & Natural Resources* 10, no. 2 (1997).

Pryne, Eric. "Government's Ax May Come Down Hard on Forks Timber Spokesman Larry Mason." *Seattle Times*, May 5, 1994.

Romano, Mike. "Who Killed the Timber Task Force?" *Seattle Weekly*, Oct. 9, 2006.

Speece, Darren Frederick. *Defending Giants: The Redwood Wars and the Transformation of American Environmental Politics*. Seattle: University of Washington Press, 2017.

Stein, Mark A. "'Redwood Summer': It Was Guerrilla Warfare: Protesters' anti-logging tactics fail to halt North Coast timber harvest. Encounters leave loggers resentful." *Los Angeles Times*, Sept. 2, 1990.

Widick, Richard. *Trouble in the Forest: California's Redwood Timber Wars*. Minneapolis: University of Minnesota Press, 2009.

CHAPTER 6: THE GATEWAY TO THE REDWOODS

Author's personal notes and photographs, Sept. 2019.

Barlow, Ron. Interview with the author, Oct. 2021.

California State Parks. "What Is Burl?" https://www.nps.gov
/redw/planyourvisit/upload/Redwood_Burl_Final-508.pdf.

Del Tredici, Peter. "Redwood Burls: Immortality Underground."
Arnoldia 59, no. 3 (1999).

Logan, William Bryant. *Sprout Lands: Tending the Endless Gift of
Trees*. New York: W. W. Norton, 2019.

Marteache, Nerea, and Stephen F. Pires. "Choice Structuring
Properties of Natural Resource Theft: An Examination of
Redwood Burl Poaching." *Deviant Behavior* 41, no. 3 (2019).

McCormick, Evelyn. *The Tall Tree Forest: A North Coast Tree
Finder*. Self-published, Rio Dell, 1987.

Peattie, Donald Culross. *A Natural History of North American
Trees*. San Antonio, TX: Trinity University Press, 2007.

Perlin, John. *A Forest Journey: The Story of Wood and Civilization*.
Woodstock, VT: The Countryman Press, 1989.

Pires, Stephen F., et al. "Redwood Burl Poaching in the Red-
wood State & National Parks, California, USA," in Lemieux,
A. M., ed., *The Poaching Diaries* (vol. 1): *Crime Scripting for
Wilderness Problems*. Phoenix: Center for Problem Oriented
Policing, Arizona State University, 2020.

Popkin, Gabriel. "'Wood wide web'—the underground network
of microbes that connects trees—mapped for first time."
Science, May 15, 2019.

Redwood National and State Parks. "Arrest Made in Burl Poaching Case." May 14, 2014. https://www.nps.gov/redw/learn
/news/arrest-made-in-burl-poaching-case.htm.

Save the Redwoods League. "Coast Redwoods." https://www
.savetheredwoods.org/redwoods/coast-redwoods/.

Sillett, Steve. Personal correspondence with the author, Oct. 2019.

Taylor, Preston. Interview with the author, Feb. 2020.

Tudge, Colin. *The Tree: A Natural History of What Trees Are, How They Live, and Why They Matter*. New York: Crown, 2006.

University of California Agriculture and Natural Resources. "Coast Redwood (Sequoia sempervirens)." https://ucanr.edu/sites /forestry/California_forests/http___ucanrorg_sites_forestry _California_forests_Tree_Identification_/Coast_Redwood _Sequoia_sempervirens_198/.

University of Delaware. "How plants protect themselves by emitting scent cues for birds." Aug. 15, 2018.

Virginia Tech, College of Natural Resources and Environment. "Fire ecology." http://dendro.cnre.vt.edu/forsite/valentine /fire_ecology.htm.

Widick, Richard. *Trouble in the Forest: California's Redwood Timber Wars*. Minneapolis: University of Minnesota Press, 2009.

Wohlleben, Peter. *The Hidden Life of Trees: What They Feel, How They Communicate—Discoveries from a Secret World*. Vancouver, BC: Greystone Books, 2016.

CHAPTER 7: TREE TROUBLES

Cook, Terry, and Cherish Guffie. Interview with the author, Sept. 2019.

Court filings, "State of Washington v. Christopher David Guffie." Case no. 94-1-00102, accessed Oct. 2020.

Court filings, "State of Washington v. Daniel Edward Garcia." Case no. 94-1-00103, accessed Oct. 2020.

Garcia, Danny. Interviews with the author, Dec. 2019, Jan.

2020, Oct. 2020, Dec. 2020, Feb. 2021, June 2021, July 2021, and Oct. 2021.

Guffie, Chris. Interviews with the author, Sept. 2019 and Sept. 2020.

Guffie, John. Interview with the author, Oct. 2020.

Obituary of Ronald Cook, *Times-Standard* (Eureka, CA), July 20, 1976.

Obituary of Thelma Cook, *Times-Standard* (Eureka, CA), Aug. 28, 2007.

Obituary of Timmy Dale Cook, *Times-Standard* (Eureka, CA), Oct. 12, 2004.

"Victim of Crash Dies." *Times-Standard* (Eureka, CA), Mar. 1, 1971.

CHAPTER 8: MUSIC WOOD

Court filings, "United States of America v. Reid Johnston." Case no. CR11-5539RJB, accessed 2014.

Cronn, Richard, et al. "Range-wide assessment of a SNP panel for individualization and geolocalization of bigleaf maple (*Acer macrophyllum* Pursh). *Forensic Science International: Animals and Environments.* Vol. 1, Nov. 2021: 100033.

Diggs, Matthew. Interview with the author, 2014.

Durkan, Jenny. "Brinnon Man Indicted for Tree Theft from Olympic National Forest." United States Attorney's Office, Western District of Washington. Nov. 10, 2011.

"Fatality accident: Brinnon's Stan Johnston dies in crash on Hwy. 101; Candy Johnston recovering at Harbourview." *Leader* (Port Townsend, WA), Feb. 19, 2011.

Greenpeace. "Taylor, Gibson, Martin and Fender Team with Greenpeace to Promote Sustainable Logging." July 6, 2010.

https://www.greenpeace.org/usa/news/taylor-gibson-martin-and-fen/

Halverson, Matthew. "Legends of the Fallen." *Seattle Met*, Apr. 2013.

Jenkins, Austin. "Music Wood Poaching Case Targets Mill Owner Who Sold to PRS Guitars." NWNewsNetwork, Aug. 6, 2015.

Minden, Anne. Interview with the author, Aug. 2018.

National Park Service. Freedom of Information Act Request, NPS-2019-01621, accessed Nov. 2019.

———. "Size of the Giant Sequoia." Feb. 2007.

———. "Two men sentenced for theft of 'music wood' timber in Olympic National Park." Feb. 16, 2018.

O'Hagan, Maureen. "Plundering of timber lucrative for thieves, a problem for state." *Seattle Times,* Feb. 24, 2013.

Peattie, Donald Culross. *A Natural History of North American Trees*. San Antonio, TX: Trinity University Press, 2007.

Riggs, Keith. "Timber thief in Washington cuts down 300-year-old tree." Forest Service Office of Communication, Jan. 10, 2013.

Taylor, Preston. Interview with the author, Feb. 2020.

Tudge, Colin. *The Tree: A Natural History of What Trees Are, How They Live, and Why They Matter*. New York: Crown, 2006.

United States Department of Agriculture, Forest Service. "Douglas-Fir: An American Wood." FS-235.

——— "Species: Pseudotsuga menziesii var. menziesii," distributed by the Fire Effects Information System. https://www.fs.fed.us/database/feis/plants/tree/psemenm/all.html.

CHAPTER 9: THE TREES OF MYSTERY

Barlow, Ron. Interview with the author, Oct. 2021.

Cook, Terry, and Cherish Guffie. Interview with the author, Sept. 2019.

Court filings, "The People of the State of California v. Danny Edward Garcia." Case no. CR1402210A, accessed Aug. 2020.

Denny, Laura. Interview with the author, Sept. 2020.

Esler, Bill. "Second Redwood Burl Poacher Sentenced." *Woodworking Network,* June 23, 2014.

"Famous Burls Are Used in Many Nations." *Humboldt Times* Centennial Issue (Eureka, CA), Feb. 8, 1954.

Garcia, Danny. Interviews with the author, Dec. 2019, Jan. 2020, Oct. 2020, Dec. 2020, Feb. 2021, June 2021, July 2021, and Oct. 2021.

Guffie, Chris. Interviews with the author, Sept. 2019 and Sept. 2020.

Hagood, Jim, and Joe Hufford. Interview with the author, Sept. 2019.

"Homeland Security Asset Report Inflames Critics." *All Things Considered,* NPR, July 12, 2006.

Logan, William Bryant. *Sprout Lands: Tending the Endless Gift of Trees.* New York: W. W. Norton, 2019.

Muth, Robert M. "The persistence of poaching in advanced industrial society: Meanings and motivations—An introductory comment." *Society & Natural Resources* 11, no. 1 (1998).

National Park Service. Freedom of Information Act Request, NPS-2019-01621, accessed Nov. 2019.

Simmons, James. Interview with the author, Sept. 2020.

Squatriglia, Chuck. "Fighting back: Park managers are cracking down on thieves stealing old-growth redwood logs." *SF Gate,* Sept. 17, 2006.

Trick, Randy J. "Interdicting Timber Theft in a Safe Place:

A Statutory Solution to the Traffic Stop Problem." *Seattle Journal of Environmental Law* 2, no. 1 (2012).

Troy, Stephen. Interviews with the author, Sept. 2019, Sept. 2020, Feb. 2021, and July 2021.

CHAPTER 10: TURNING

Amador, Don. "2001 Orick Freedom Rally and Protest Update." Blue Ribbon Coalition, June 26, 2001.

Author's personal notes and photographs.

Barlow, Ron. Interview with the author, Oct. 2021.

Cart, Julie. "Storm over North Coast rights." *Los Angeles Times*, Dec. 18, 2006.

Cook, Terry, and Cherish Guffie. Interview with the author, Sept. 2019.

Court records, "California Department of Parks and Recreation v. Edward Salsedo." Case no. A112125, July 2009, accessed Jan. 2020.

Frick, Steve. Interview with the author, Sept. 2019.

Hagood, Jim. Interviews with the author, Sept. 2019 and Jan. 2021.

House, Rachelle. "Western Snowy Plover reaches important milestone in its recovery." *Audubon,* Aug. 2018.

Hughes, Derek. Interviews with the author, Sept. 2020, Oct. 2020, Mar. 2021, Apr. 2021, July 2021, and Oct. 2021.

Lehman, Jacob. "Gates draw anger." *Times-Standard* (Eureka, CA), Aug. 2000.

Meyer, Betty. Interview with the author, Sept. 2019.

Netz, Lynne. Interview with the author, Sept. 2019.

"Orick Under Siege." Advertisement. *Times-Standard* (Eureka, CA), July 29, 2000.

Pero, Branden. Interviews with the author, Sept. 2019, Sept. 2021, and Oct. 2021.

Simmons, James. Interview with the author, Sept. 2019.

Treasure, James. "'Orick in grave need,' according to letter." *Times-Standard* (Eureka, CA), Oct. 24, 2001.

Walters, Heidi. "Orick or bust." *North Coast Journal of Politics, People & Art* (Eureka, CA), May 31, 2007.

CHAPTER 11: BAD JOBS

"Adverse Community Experiences and Resilience: A Framework for Addressing and Preventing Community Trauma." Prevention Institute, 2015.

Bradel, Alejandro, and Brian Greaney. "Exploring the Link Between Drug Use and Job Status in the U.S." Federal Reserve Bank, July 2013.

Case, Anne, and Angus Deaton. *Deaths of Despair and the Future of Capitalism.* Princeton, NJ: Princeton University Press, 2021.

"Coley." *Intervention,* Season 3, Episode 11. A&E, aired Aug. 2007.

Coriel, Andrew, and Phil Huff. Interview with the author, July 2020.

Court filings, "The People of the State of California v. Danny Edward Garcia." Case no. CR1402210A, accessed Aug. 2020.

Daniulaityte, Raminta, et al. "Methamphetamine Use and Its Correlates among Individuals with Opioid Use Disorder in a Midwestern U.S. City." *Substance use & misuse* 55, no. 11 (2020): 1781–1789.

DataUSA. "Orick, CA." https://datausa.io/profile/geo/orick-ca/.

Dumont, Clayton W. "The Demise of Community and Ecology in the Pacific Northwest: Historical Roots of the Ancient

Forest Conflict." *Sociological Perspectives* 39, no. 2 (1996): 277–300.

Goldsby, Mike. Interview with the author, Sept. 2019.

Guffie, Chris. Interview with the author, Sept. 2020.

Hagood, Jim. Interviews with the author, Sept. 2019 and Jan. 2021.

Heffernan, Virginia. "Confronting a Crystal Meth Head Who Is Handy with a Chainsaw." *New York Times,* Aug. 10, 2007.

Henkel, Dieter. "Unemployment and substance use: A review of the literature (1990–2010)." *Current Drug Abuse Reviews* 4, no. 1 (2011).

Hufford, Donna, and Joe Hufford. Interview with the author, Sept. 2019.

Hughes, Derek. Interviews with the author, Sept. 2020, Oct. 2020, Mar. 2021, Apr. 2021, July 2021, and Oct. 2021.

"Humboldt County Economic & Demographic Profile." Center for Economic Development, 2018.

Kemp, Kym. "Never Ask What a Humboldter Does for a Living and Other Unique Etiquette Rules." *Lost Post Outpost* (Eureka, CA), Jan. 8, 2011.

Kristof, Nicholas D., and Sheryl WuDunn. *Tightrope: Americans Reaching for Hope.* New York: Knopf, 2020.

Life After Meth: Facing the Northcoast Methamphetamine Crisis. Produced by Seth Frankel and Claire Reynolds. Eureka, CA: KEET-TV, 2006.

Lupick, Travis. *Fighting for Space: How a Group of Drug Users Transformed One City's Struggle with Addiction.* Vancouver, BC: Arsenal Pulp Press, 2017.

Madonia, Joseph F. "The Trauma of Unemployment and Its Consequences." *Social Casework* 64, no. 8 (1983): 482–88.

Maté, Gabor. *In the Realm of Hungry Ghosts: Close Encounters with Addiction.* Toronto: Random House Canada, 2009.

"Methamphetamine." California Northern and Eastern Districts Drug Threat Assessment, National Drug Intelligence Center, Jan. 2001.

Minden, Anne. Interview with the author, Aug. 2018.

Robles, Frances. "Meth, the Forgotten Killer, Is Back. And It's Everywhere." *New York Times*, Feb. 13, 2018.

Rose, David. "'The Pacific Northwest is drowning in methamphetamine': 17 arrested in major drug trafficking operation." Fox13 Seattle, Oct. 24, 2019.

Sherman, Jennifer. "Bend to Avoid Breaking: Job Loss, Gender Norms, and Family Stability in Rural America." *Social Problems* 56, no. 4 (2009).

———. *Those Who Work, Those Who Don't: Poverty, Morality, and Family in Rural America.* Minneapolis: University of Minnesota Press, 2009.

Trick, Randy J. "Interdicting Timber Theft in a Safe Space: A Statutory Solution to the Traffic Stop Problem." *Seattle Journal of Environmental Law* 2, no. 1 (2012): 383–426.

Volkow, Dr. Nora. "Rising Stimulant Deaths Show That We Face More Than Just an Opioid Crisis." National Institute on Drug Abuse, Nov. 2020.

Widick, Richard. *Trouble in the Forest: California's Redwood Timber Wars.* Minneapolis: University of Minnesota Press, 2009.

Yu, Steve. Interview with the author, July 2020.

CHAPTER 12: CATCHING AN OUTLAW

"Arrest made in burl poaching case." Redwood National and State Parks, May 14, 2014.

Author's personal notes and photographs.

Brown, Patricia Leigh. "Poachers Attack Beloved Elders of California, Its Redwoods." *New York Times*, Apr. 8, 2014.

Cook, Terry, and Cherish Guffie. Interview with the author, Sept. 2019.

Court filings, "The People of the State of California v. Danny Edward Garcia." Case no. CR1402210A, accessed Aug. 2020.

Pires, Stephen F., et al. "Redwood Burl Poaching in the Redwood State & National Parks, California, USA," in Lemieux, A. M., ed., *The Poaching Diaries* (vol. 1): *Crime Scripting for Wilderness Problems.* Phoenix: Center for Problem Oriented Policing, Arizona State University, 2020.

Simon, Melissa. "Burl poacher sentenced to community service." *Times-Standard* (Eureka, CA), June 20, 2014.

Sims, Hank. "Burl Poaching Suspect Arrested." *Lost Coast Outpost* (Eureka, CA), May 14, 2014.

Yu, Steve. Interview with the author, July 2020.

CHAPTER 13: IN THE BLOCKS

Author's personal notes and photographs.

Cook, Terry, and Cherish Guffie. Interview with the author, Sept. 2019.

Court filings, "People of the State of California v. Derek Alwin Hughes." Case no. CR1803044, accessed Dec. 2020.

"The Dangers of Being a Ranger." *Weekend Edition,* NPR, June 18, 2005.

Davidson, Joe. "Federal land employees were threatened or assaulted 360 times in recent years, GAO says." *Washington Post,* Oct. 21, 2019.

Garcia, Danny. Interviews with the author, Dec. 2019, Jan.

2020, Oct. 2020, Dec. 2020, Feb. 2021, June 2021, July 2021, and Oct. 2021.

Hearne, Rick. "Figuring out figure—bird's eye." *Wood Magazine*. https://www.woodmagazine.com/materials-guide/lumber/wood-figure/figuring-out-figure—birds-eye.

Hughes, Derek. Interviews with the author, Sept. 2020, Oct. 2020, Mar. 2021, Apr. 2021, July 2021, and Oct. 2021.

Johnson, Kirk. "In the Wild, a Big Threat to Rangers: Humans." *New York Times*, Dec. 6, 2010.

Netz, Lynne. Interview with the author, Sept. 2019.

Pennaz, Alice B. Kelly. "Is That Gun for the Bears? The National Park Service Ranger as a Historically Contradictory Figure." *Conservation & Society* 15, no. 3 (2017): 243–54.

Pero, Branden. Interviews with the author, Sept. 2019, Sept. 2021, and Oct. 2021.

Probation Report, "The People of the State of California v. Derek Alwin Hughes." Aug. 2021, accessed Oct. 2021.

Trick, Randy J. "Interdicting Timber Theft in a Safe Space: A Statutory Solution to the Traffic Stop Problem." *Seattle Journal of Environmental Law* 2, no. 1 (2012): 383–426.

Troy, Stephen. Interviews with the author, Sept. 2019, Sept. 2020, Feb. 2021, July 2021, and Oct. 2021.

CHAPTER 14: PUZZLE PIECES

Barnard, Jeff. "Redwood park closes road to deter burl poachers." Associated Press, Mar. 5, 2014.

Court filings, "People of the State of California v. Derek Alwin Hughes." Case no. CR1803044, accessed Dec. 2020.

Hughes, Derek. Interviews with the author, Sept. 2020, Oct. 2020, Mar. 2021, Apr. 2021, July 2021, and Oct. 2021.

Pero, Branden. Interviews with the author, Sept. 2019, Sept. 2021, and Oct. 2021.

Probation Report, "The People of the State of California v. Derek Alwin Hughes." Aug. 2021, accessed Oct. 2021.

Sims, Hank. "Humboldt Deputy DA Named California's 'Wildlife Prosecutor of the Year': Kamada Prosecuted Poachers, Growers, Dudleya Bandits." *North Coast Outpost* (Eureka, CA), June 21, 2018.

Troy, Stephen. Interviews with the author, Sept. 2019, Sept. 2020, Feb. 2021, July 2021, and Oct. 2021.

CHAPTER 15: A NEW SURGE

British Columbia Ministry of Forests, Lands and Natural Resource Operations. "Tree poaching—response provided Oct. 2018." Personal correspondence with the author, Feb. 2019.

———. "Unauthorized Harvest Statistics: 2016–2018." Personal correspondence with the author, Feb. 2019.

Clarke, Luke. Interview with the author, Mar. 2019.

"Forest Stewardship Plan." Sunshine Coast Community Forest. http://www.sccf.ca/forest-stewardship/forest-stewardship-plan, accessed Aug. 19, 2021.

Holt, Rachel, et al. "Defining old growth and recovering old growth on the coast: Discussion of options." Prepared for the Ecosystem Based Management Working Group, Sept. 2008.

Hooper, Tyler (Canada Border Services Agency). Personal correspondence with the author, Apr. 2021.

Lasser, Dave. Interview with the author, Sept. 2020.

Nanaimo Homeless Coalition. "Factsheet: Homelessness in Nanaimo," 2019.

Peterson, Jodi. "Northwest timber poaching increases." *High Country News* (Paonia, CO), June 8, 2018.

"Story of the year: DisconTent City." *Nanaimo (BC) News Bulletin,* Dec. 27, 2018.

Sunshine Coast Community Forest. "History." http://www.sccf .ca/who-we-are/history.

"Timber poaching grows on Washington public land." Washington Forest Protection Association Blog, Dec. 19, 2018. https://www.wfpa.org/news-resources/blog/timber -poaching-grows-on-washington-public-land/.

"Tree poaching hits 'epidemic' levels." *Coast Reporter* (Sechelt, BC), May 18, 2020.

Vinh, Pamela. Interview with the author, Feb. 2019.

Washington Department of Natural Resources. "Economic & Revenue Forecast," Feb. 2018.

Zeidler, Maryse. "Report recommends batons, pepper spray for B.C. natural resource officers." CBC.ca, Mar. 10, 2019.

Zieleman, Sara. Personal correspondence with the author, Apr. 2021.

CHAPTER 16: THE ORIGIN TREE

Court filings, "United States of America v. Justin Andrew Wilke." Case no. CR19-5364BHS, accessed Sept. 2021.

Golden, Hallie. "'A problem in every national forest': Tree thieves were behind Washington wildfire." *Guardian* (London), Oct. 5, 2019.

"Member of timber poaching group that set Olympic National Forest wildfire sentenced to 2½ years in prison." United States Attorney's Office, Western District of Washington, Sept. 21, 2020.

United States Department of Agriculture, Forest Service. "Maple Fire investigation results," Oct. 1, 2019.

CHAPTER 17: TRACKING TIMBER

Adventure Scientists. "Timber Tracking." https://www.adventure scientists.org/timber.html.

———. "Tree DNA Used to Convict Timber Poacher," July 29, 2021.

Cronn, Richard. Interview with the author, Aug. 2021.

Cronn, Richard, et al. "Range-wide assessment of a SNP panel for individualization and geolocalization of bigleaf maple (*Acer macrophyllum* Pursh). *Forensic Science International: Animals and Environments*. Vol. 1, Nov. 2021: 100033.

Dowling, Michelle, Michelle Toshack, and Maris Fessenden. "Timber Project Report 2019." Adventure Scientists, Nov. 2020. https://www.adventurescientists.org/uploads/7/3/9/8/7398741/2019_timber-report_20201112.pdf.

Gupta, P., J. Roy, and M. Prasad. "Single nucleotide polymorphisms: A new paradigm for molecular marker technology and DNA polymorphism detection with emphasis on their use in plants." *Current Science* 80, no. 4 (Feb. 2001): 524–35.

United States Department of Agriculture, Forest Service. "Maple Fire investigation results," Oct. 1, 2019.

CHAPTER 18: "IT WAS A VISION QUEST"

Author's personal notes and photographs, Sept. 2019.

Baquero, Diego Cazar. "Indigenous Amazonian communities

bear the burden of Ecuador's balsa boom." Mongabay.com, Aug. 17, 2021.

Davidson, Helen. "From a forest in Papua New Guinea to a floor in Sydney: How China is getting rich off Pacific lumber." *Guardian* (London), May 31, 2021.

Dunlevie, James. "Million-dollar 'firewood theft' operation busted in southern Tasmania." ABC News (Sydney), May 7, 2020.

Espinoza, Ed. Interviews with the author, June 2018 and Sept. 2019.

Food and Agriculture Organization of the United Nations. North American Forest Commission, Twentieth Session, "State of Forestry in the United States of America," 2000. http://www.fao.org/3/x4995e/x4995e.htm.

Goddard, Ken. Interviews with the author, June 2018 and Sept. 2019.

Grant, Jason, and Hin Keong Chen. "Using Wood Forensic Science to Deter Corruption and Illegality in the Timber Trade." Targeting Natural Resource Corruption (Topic Brief), Mar. 2021.

Grosz, Terry. Interview with the author, June 2018.

International Bank for Reconstruction and Development/The World Bank. *Illegal Logging, Fishing, and Wildlife Trade: The Costs and How to Combat It.* Oct. 2019.

Lancaster, Cady. Interviews with the author, Sept. 2019 and Oct. 2020.

Mukpo, Ashoka. "Ikea using illegally sourced wood from Ukraine, campaigners say." Mongabay.com, June 29, 2020.

Nellemann, Christian. Interview with the author, Sept. 2013.

Neme, Laurel A. *Animal Investigators: How the World's First Wildlife Forensics Lab Is Solving Crimes and Saving Endangered Species.* New York: Scribner, 2009.

Petrich, Katharine. "Cows, Charcoal, and Cocaine: al-Shabab's Criminal Activities in the Horn of Africa." *Studies in Conflict & Terrorism*, 2019.

Sheikh, Pervaze A. "Illegal Logging: Background and Issues." Congressional Research Service, June 2008. https://crsreports.congress.gov/product/pdf/RL/RL33932/8.

World Wide Fund for Nature. "Illegal wood for the European market," July 2008.

———. "Stop Illegal Logging." https://www.worldwildlife.org/initiatives/stopping-illegal-logging.

Zuckerman, Jocelyn C. "The Time Has Come to Rein in the Global Scourge of Palm Oil." *Yale Environment 360*, May 27, 2021.

CHAPTER 19: FROM PERU TO HOUSTON

Aguirre, Ruhiler. Interviews with the author, Apr. 2018.

Author's personal notes and photographs.

Conniff, Richard. "Chasing the Illegal Loggers Looting the Amazon Forest." *Wired*, Oct. 2017.

Custodio, Leslie Moreno. "In the Peruvian Amazon, the prized shihuahuaco tree faces a grim future." Mongabay.com, Oct. 31, 2018.

Environmental Investigation Agency. "The Illegal Logging Crisis in Honduras," 2006.

———. "The Laundering Machine: How Fraud and Corruption in Peru's Concession System Are Destroying the Future of Its Forests," 2012.

Urrunaga, Julia. Interview with the author, May 2018.

CHAPTER 20: "WE TRUST IN THE TREES"

Author's personal notes and photographs.

Jumanga, Jose. Interview with the author, May 2018.

Vasquez, Raul. Interview with the author, May 2018.

CHAPTER 21: CARBON SINKS

Author's personal notes and photographs.

Ennes, Juliana. "Illegal logging reaches Amazon's untouched core, 'terrifying' research shows." Mongabay.com, Sept. 15, 2021.

Espinoza, Ed. Interviews with the author, June 2018 and Sept. 2019.

Carrington, Damian. "Amazon rainforest now emitting more CO_2 than it absorbs." *Guardian* (London), July 14, 2021.

Center for Climate and Energy Solutions. "Wildfires and Climate Change." https://www.c2es.org/content/wildfires-and-climate-change/.

International Union for Conservation of Nature. "Peatlands and climate change." Issues Brief, 2014.

———. "Rising murder toll of park rangers calls for tougher laws." July 29, 2014.

Jirenuwat, Ryn, and Tyler Roney. "The guardians of Siamese rosewood." China Dialogue.net, Jan. 28, 2021.

Lancaster, Cady. Interviews with the author, Sept. 2019 and Oct. 2020.

Law, Beverly, and William Moomaw. "Curb climate change the easy way: Don't cut down big trees." The Conversation.com, Apr. 7, 2021.

Rainforest Alliance. "Spatial data requirements and guidance," June 2018.

Shukman, David. "'Football pitch' of Amazon forest lost every minute." BBC News, July 2, 2019.

United Nations Sustainable Development. "UN Report: Nature's Dangerous Decline 'Unprecedented'; Species Extinction Rates 'Accelerating.'" May 6, 2019.

Young, Rory, and Yakov Alekseyev. "A Field Manual for Anti-Poaching Activities." African Lion & Environmental Research Trust, 2014.

CHAPTER 22: IN LIMBO

Barlow, Ron. Interview with the author, Oct. 2021.

Cook, Terry, and Cherish Guffie. Interview with the author, Sept. 2019.

Garcia, Danny. Interviews with the author, Dec. 2019, Jan. 2020, Oct. 2020, Dec. 2020, Feb. 2021, June 2021, July 2021, and Oct. 2021.

Guffie, John. Interview with the author, Oct. 2020.

Hughes, Derek. Interviews with the author, Sept. 2020, Oct. 2020, Mar. 2021, Apr. 2021, July 2021, and Oct. 2021.

Pero, Branden. Interviews with the author, Sept. 2019 and Sept. 2021.

Probation Report, "The People of the State of California v. Derek Alwin Hughes." Aug. 2021, accessed Oct. 2021.

AFTERWORD

Bray, David. "Mexican communities manage their local forests, generating benefits for humans, trees and wildlife." The Conversation.com. https://theconversation.com/mexican

-communities-manage-their-local-forests-generating-benefits
-for-humans-trees-and-wildlife-165647.

British Columbia Community Forest Association. "Community
Forest Indicators 2021," Sept. 2021.

Duffy, Rosaleen, et al. "Open Letter to the Lead Authors
of 'Protecting 30% of the Planet for Nature: Costs,
Benefits and Implications.'" https://openlettertowaldronetal
.wordpress.com/.

Meissner, Dirk. "Ongoing protests, arrests at Fairy Creek over logging
'not working,' says judge." Canadian Press, Sept. 18, 2021.

Polmateer, Jaime. "172 job layoffs as Canfor announces closure
of Vavenby mill." *Clearwater (BC) Times,* June 3, 2019.

Waldron, Anthony, et al. "Protecting 30% of the Planet for Nature: Costs, Benefits and Economic Implications." Working
paper analyzing the economic implications of the proposed
30% target for areal protection in the draft post-2020 Global
Biodiversity Framework. Cambridge Conservation Research
Institute, 2020.

ARCHIVES

Clarke Historical Museum, interpretive galleries

Humboldt County Historical Society (HCHS), archival collections

The New York Public Library, Digital Map Collections

University of California Berkeley: Bancroft Library, Oral History Center. Transcripts of interviews conducted by Amelia Fry.

INDEX

ABOUT THE AUTHOR

Lyndsie Bourgon is a writer, an oral historian, and a 2018 National Geographic Explorer based in British Columbia. She writes about the environment and its intersections with history, culture, and identity. Her features have been published in *The Atlantic, Smithsonian, The Guardian,* the *Oxford American, Aeon, The Walrus,* and *Hazlitt,* among other outlets.

FATHER FIGURE

Are You in the House Alone?
The Ghost Belonged to Me
Ghosts I Have Been
Representing Super Doll
Through a Brief Darkness

THE VIKING PRESS, NEW YORK

FATHER FIGURE

a novel by

RICHARD PECK

G/BR

First Edition
Copyright © Richard Peck, 1978
All rights reserved
First published in 1978 by The Viking Press
625 Madison Avenue, New York, N.Y. 10022
Published simultaneously in Canada by
Penguin Books Canada Limited
Printed in U.S.A.
1 2 3 4 5 82 81 80 79 78

Library of Congress Cataloging in Publication Data
Peck, Richard. Father figure.
Summary: After being a father figure for years, Jim
and his young brother are reunited with their divorced
father and Jim is forced to find a new role for himself.
[1. Brothers and sisters—Fiction. 2. Fathers
and sons—Fiction] I. Title.
PZ7.P338Fat   [Fic]   78-7909
ISBN 0-670-30930-3

This book is for
Robert Unsworth
George Nicholson
Richard Brundage
and for
Madeline and Sam Paetro

FATHER FIGURE

ONE

When my mom died, I spent a lot of time—maybe too much—trying to visualize how it happened. My kid brother, Byron, was going over it in his mind too, in his own way. Even at a distance I like to believe I can hear him thinking, and I thought I ought to be ready if he ever asked me anything.

When somebody dies alone, you try to fill in the details, maybe to make up for not being there and changing the whole history of the thing. It's one of those times you want to be able to account for. An alibi for some higher authority than the police: *James Atwater, where were you on the night of April twenty-ninth between the hours of* . . .

It was a Tuesday night, symphony night for my grandmother. She'd gone to the Brooklyn Academy of Music with her friend, Mrs. Schermerhorn. They've got season tickets, and they've sat in the same two seats every week for maybe thirty-five, forty years. Mrs. Schermerhorn lives on Joralemon Street, directly behind us. Her old stretched Cadillac Fleetwood is garaged in the alley between. Her garage doors face ours.

I can picture Nathan, her chauffeur, nosing the old Fleetwood out of the alley that Tuesday evening, braking at the corner, the red taillights turning the alley red. Swinging around to Joralemon Street to pick up Mrs. Schermerhorn on her front steps. Then doubling back and creeping up Remsen Street, where my grandmother always waits in the bay window, leaning on her Lucite cane and looking at her watch.

That night my brother, Byron, had gone with a bunch of kids from his school and somebody's father to the Botanic Garden to see a display of bonsai trees, those little stunted trees the Japanese train down to potted plants, miniature versions of the real thing. Byron's not particularly into stunted trees, but anything interests him. At the age of eight he's picked up weird snatches of higher learning from somewhere. Where, I couldn't say. That isn't the way my mind works.

I was all the way up in the Bronx in Van Cortlandt Park that night. It's down the hill from my school, the Van Cortlandt Academy for Boys. Our playing fields are across Broadway at the edge of the park. There's a hill farther in, where one of Washington's generals lighted signal fires during the Revolutionary War. The place is full of cross-country runners, West Indians playing cricket on their own pitches, and our ball diamonds.

I'm kind of sub-coach for the Lower School ball team. Not what you'd call a team. Put it this way: I'm father figure for our own Bad News Bears. We practiced late that night because the floods were on in the park in spite of the energy crisis. The little guys thought being out this late was a big deal. I never pushed athletics with Byron, who doesn't even go to my school. What he wants he takes on his own, and you can push a kid only so hard.

I had a couple fourth graders off to one side swinging leaded bats for muscle building, and a double row of fifth graders organized to play pepper. Then I gave my sixth graders a little running commentary on the rock-bottom basics, like don't bunt into short grass, and try to hit to the right if there's a man on second. We don't get much past the fundamentals. Not with a self-taught coach who learned to slide by watching the ground come up at him. I don't know if they learn anything, but they like the attention.

When it got really dark, some of the younger ones started glancing up at the sky and out past the lights. So I yelled for them to wrap it up and get all the equipment together. I wear this chrome whistle on a lanyard the kids gave me for a present, but I never use it. I'm not that much of a coach type. So I just yell a lot. But they like to see me wearing the whistle as a badge. They grouped up in a convoy behind the two biggest sixth graders for the walk to the subway. They're committed to their tough image, but none of them wants to tackle the subway alone at that age. I wouldn't want them to, anyway.

I stayed on in the park by myself, running the bases a few times, practicing my hook slide which is reasonably dependable but nothing to look at. Then it got too dark even for me. You could see the white legs of people pumping along the park paths in the distance, but in New York you don't know if they're jogging for health or running for help. I took the subway home, grabbed something to eat out of the refrigerator since Almah wasn't on duty in the kitchen, and went up to bed. Probably I figured my mom was already asleep. She needed a lot of rest by then.

So once I account for everybody else, I have to think about how Mom died. That's what's left. She died in her car. A Buick Skylark, not new. I can almost see her in the driver's seat, with one hand at the bottom of the steering wheel, kind of resting there the way she always drove, and the seat-belt strap pulled tight over her left shoulder. It's hot in the car and getting stuffier. There's an arc of beaded haze clouding up the rear window and getting bigger. Hanging down from the ignition key is the ring with the silver heart Byron and I gave her last

Easter. The New York State inspection sticker's peeling on the windshield.

Outside the car it's completely dark. But still, Mom sees pinwheels and zigzags of light in front of the hood. The air's heavy and sweet, like up in the country, where the roads tunnel through the trees. It's like not being outdoors at all. There's the sudden light from fireflies and the glowing eyes of little animals crouching on the center stripe, hypnotized by surprise, until the last second when they make a run for the ditch.

I feel the tiredness creeping up on Mom, fogging her thoughts and her eyesight, making the dark brighter. She takes her hand off the wheel for a moment to scoop the strands of hair back from her forehead the way she always did when she was really worn out. I can see her slipping past thinking and maybe dreaming something meaningless, the way you do sometimes when you're really still awake. I can imagine her not having any pain or fear. Then between one moment and the next: nothing.

When Byron came home that night, I suppose he checked to see if I was in bed. I think I remember the door opening and closing. He usually made the rounds at night, looking for his cat, Nub, using this as a standard excuse to look in on me. Even though he knew I always threw Nub out of my room before I went to bed because he'd sleep on your face if he got the chance. Also Nub sometimes brought in a little tribute and left it on your

[7]

pillow: a dead mouse, an exhausted roach, a moth missing its wings—something like that.

Grandmother came in about eleven as usual, though I
was completely out by then and didn't hear her cane on
the stairs. It was Mrs. Schermerhorn's chauffeur, Nathan,
who found Mom early the next morning, the last day of
April. He was out in the alley by six, wiping down the
Cadillac with a KozaK cloth. Nathan's pretty much in
charge of the alley, always going around to see that
everybody's garbage can has a tight lid, and keeping a
ring of all the garage keys for people who are always losing their own.

He must have seen our garage windows were fogged
over. By then the Buick had run out of gas, so he
wouldn't have heard the engine idling. Anyway, he must
have worked through his ring of keys till he came to ours
—he color-codes them. Then when he swung our garage
door up he must have seen the hose running from the
car's exhaust and snaking up into the wing window. He
found Mom sitting upright in the front seat, where she'd
killed herself.

I can picture Nathan jerking at the door handle and
finding it locked. I can even see his old fist wrapped in
the KozaK cloth shattering the driver's side window with
one blow. His arm with the snow-white cuff turned back
is angling inside through the broken star of glass to ease
the door lock up. And then he yanks the door open,
holding his breath against the carbon monoxide. He

[8]

reaches out to Mom still held in place by the shoulder-strap seat belt, with her head resting back. I can see Nathan's hand reaching for Mom's wrist, even though he'd know there wasn't any life left in her.

TWO

All the details of death on TV shows are lies. All those beautiful female mourners with short black skirts and great legs and dark glasses. All those silent Mercedeses winding through acid-green graveyards in a fine mist. The pin-striped undertakers. The organ music outdoors. And no real grief.

The night before Mom's funeral, friends are invited to call at Loring and Sons' funeral establishment in Manhattan, a high-rise mortuary with a different deceased on every level and people in the elevators calling out the names of the dead to be sure they get the right floor. In a way Loring's is a class act, not like the Brooklyn funeral parlors along Atlantic Avenue where you can see the body from the front door.

[10]

I spend most of two days feeling fairly numb, trying not to go too near myself, trying to keep Byron in sight without hanging over him. But he isn't saying anything. And I don't know how to begin with him. He isn't doing anything either, and usually he's got a dozen projects lined up and going.

The guy in charge of the evening at Loring's has his own image: no pinstripes or white hands. He's trim looking, in a blue suit from Barney's, with wide lapels. I get weird vibrations off people that night. I think maybe he's just jogged over from a couple of fast tennis sets at the Grand Central Racquet Club. His hands are square, and he has a firm, dry handshake.

When we get there in Mrs. Schermerhorn's car, people are already signing the guest register. They stand around in knots ankle-deep in the carpet, glancing over at Grandmother and Mrs. Schermerhorn and Byron and me. Mom's coffin is between two floor lamps with torch-shaped glass shades. The lid's closed, according to Grandmother's instructions. Since there's nothing to see, people stay away from it. Grandmother's thought that through in advance.

All the heads in the room are half bowed. Some of the men hold their chins in their hands. The women brush things off their sleeves, and nobody speaks above a low hum. They hold back from coming over to us, nobody wanting to be first.

I watch how Grandmother grips the crook of her

cane and tries not to lean on Mrs. Schermerhorn while they move very slowly across the room.

You don't second-guess Grandmother. I don't know what she's thinking. The biggest irony in the world must be to lose your own child, your only one. Maybe Grandmother thought she was the one who should be in that closed coffin. Maybe she thought there'd been some vast cosmic mix-up. And maybe there had. Whatever she was thinking, she was thinking it with a clear head. Dr. Painter had come around to the house that afternoon, bringing a bottle of Valium. But I don't think Grandmother took any. It's hard enough for her to accept a cane. She's not about to give in to a crutch.

Byron and I hang around the door while she and Mrs. Schermerhorn head for a sofa in the corner away from the coffin. Their heads are close together: Grandmother's nearly white, with every hair in place; Mrs. Schermerhorn's hair mahogany-dyed, springy and thin. Then Grandmother settles onto the sofa, giving her cane the usual little wrist action to scoot it behind her heels, where people won't fall over it or even notice it.

"Would you and your little brother like to . . . step up . . . nearer your mother?" I wonder if the undertaker in the Barney's suit is new on the job. He doesn't seem to have it together. I look down at Byron. It's a long way. I'm five eleven, practically, and he hasn't grown a quarter of an inch in six months. He's wearing a tie I knotted for him. No coat, and the tie's too long for him. It laps down over his fly. He doesn't seem to be noticing anything much, so I nod to the undertaker, and Byron and I walk down the center of the room to where Mom is. The undertaker keeps a half step behind us, and people make way.

When we get to the coffin it's like standing on the rim of a cliff. We've gone as far as we can go. Polished bronze, and a big spray of white rosebuds without a card or ribbon or anything. I think Grandmother asked Mrs. Schermerhorn to order the flowers. But there's nothing real about any of this scene.

Byron must be thinking the same thing. He speaks for the first time all day. I barely hear him. "Is she in there, Jim?"

I knew whenever he finally said something, I wouldn't be ready for it. "Yes, Mom's in there, in a way. In another way, she isn't."

"I don't mean the difference between the body and the soul," he says patiently. "I mean, is she cremated yet or not?"

"Not yet. Tomorrow, after the funeral."

I can feel people's eyes on us. Since our backs are to them, they're free to look. The dead woman's two sons paying their last respects. I never know what that saying means: paying your last respects. My hand automatically goes to my back pocket to see if I've remembered to bring my billfold.

This doesn't make any sense, so I think about trying

to keep the communication open with Byron.

"She was dying anyway, By. You knew that. Nobody made a big secret of it."

"I know," he says. "She was hurting bad. Especially in the night."

I really need to cry now. A simple biological need. I'm too old just to stand there looking solemn like Byron. I have other—needs. I can't let him see me falling apart. Still, it's beginning to happen. SON COLLAPSES IN GRIEF ACROSS SUICIDE MOTHER'S COFFIN. Headlines blare inside my head. I can feel my face crumpling, from the chin up, and my throat closing. The undertaker's hand falls on my shoulder, resting there without pressure. And this nearly pushes me over the edge.

But help's on the way. The white rosebuds on Mom's coffin are beginning to run together when a woman steps up, invading the neutral circle around Byron and me. I can't be sure, but I think she waves the undertaker away. His hand leaves my shoulder.

"You're Jim Atwater?" she says. I look at her, blinking like crazy. Byron peers around me. She's familiar in a way, some friend of Mom's who used to drop in occasionally. Not anybody too close. I think once when Mom had just come from the hospital this woman stopped by and left a stack of magazines or something.

"Yes, I'm Jim. This is my brother, Byron."

"I'm Winifred Highsmith. I went to school with your mother. You know, Katharine Gibbs, the place where

[14]

they taught us to wear white gloves and type."

Byron's giving her his total attention. Her voice has a cutting edge. But other people in the room start talking a little louder. Since somebody's made the first move toward Byron and me and the coffin, the decibel level rises.

She must be Mom's age: forty-six. She looks it. Tired New York eyes, jet-black hair pulled back tight, drawing her eyebrows up. She's nearly as tall as I am, and her eyes make contact with mine and won't let go. "I don't suppose they allow smoking in here."

"I don't know." My throat seems to be opening up again.

"It doesn't matter. It wouldn't help." She shifts the long strap of a purse higher on her shoulder. "Listen," she says, "I forgot you were going to be so grown-up. I never can keep track of people's children. One day they're wearing Pampers; the next day they're flunking out of college. You know what I mean?"

"Yeah, I think so."

"I'm not married. Never was. No kids. I always liked your mother." She talks in a jerky way, and her hand flips at the catch on her purse, going for a cigarette that wouldn't help, I guess. "I won't say all those things you're about to hear from other people. I'm rotten at that. I'll just talk a couple minutes and then leave. Okay?"

"Sure," I say. "That's—fine." I can see her clearer. My

eyes aren't blurred. She looks kind of fierce.

"Where's your sister? Aren't there three of you?"

"She can't come. She's married and living in Germany. Her husband's in the Air Force."

"*Married?* Oh, well. I suppose I must have known that. What is she, about twenty?"

"Twenty-three."

"Ye gods."

"And she couldn't travel. She's pregnant."

"How far along?"

"Seventh month."

"So you and Byron will be uncles."

"Yeah. I hadn't thought about that. I guess we will."

"Let's see. Her name's Lorraine, right? Did she marry a nice guy?"

"Better not ask Grandmother, but yes, she did."

"Not up to the mark with Granny?" Winifred Highsmith runs her tongue along the inside of her cheek.

"Not quite. He's a sergeant, and Grandmother thought the least Lorraine could do was aim herself at the officer class." I'm beginning to forget we're standing two feet from my mother's coffin, from my mother. It's like a regular conversation. I'm probably even hanging too loose.

"Your mother wouldn't have given two hoots about that kind of thing. That's one of the things I liked about her. I suppose your father isn't here."

"He—he may be here tomorrow," I say fast, wanting

[16]

to keep off the subject of Dad with a stranger while Byron's there at my elbow, all ears.

"Oh, yes." Winifred Highsmith rolls her eyes. "That's your father to a T. A dollar down and a day late, as they say."

"He'll probably be here," I say in a low voice, feeling stupid for trying to cover for him.

"But where was he when she needed him?" she says, looking fiercer. "How long was she sick—really sick? A year? Longer? Look, I'm sorry. Forget it. Howard's your father, and I should keep my big mouth shut. I never really knew the guy anyway. People's husbands— who can keep track?"

"I don't remember him that well myself. It's been about eight years or so since he left." It was exactly eight years. I know because Byron was only a baby at the time.

"You don't see him?" Winifred says.

"He lives down in Florida. It's a long way." There I go again, covering for this crumb.

"Yes," she says. "Hell of a long way. No planes, no phones, no word sent or received. Supplies sent in annually by native bearers. There, see? I can't keep still. Forget it, will you? Your mother was better off without him. Besides, she had you. Look, I've got to go now. I can take just so much of this type atmosphere. I came because I read about Barbara—your mother—in the *Times*. She was a great gal, you know what I mean? A little under your grandmother's thumb, if you know what I mean, but then who wouldn't be? And I really liked her—your mother, I mean. How many people do you really like in a lifetime?"

I shrug, not knowing. "Okay, I'm going to cut out now." She hitches the strap up on her shoulder again and opens the catch on her purse. "I'm truly not gifted at things like this. I probably won't see you again, either of you. I mean like day after tomorrow I'll pass you on the street, and you'll be married with five kids, and it'll be 1999. You know what I'm saying? I never can keep up with people's kids. You'll be fine now. In situations like this, it's getting through the first part that counts. The trouble is, I never can stick it out past that. Here come the rest of them."

Then she's gone, taking long strides across the room. She has thick calves, like a dancer. And she doesn't slow down going past Grandmother.

The rest of them move in, advancing on Byron and me from three sides. My throat tenses up again. But I can control it. Winifred Highsmith's way out in the hall now, waiting for the elevator, fumbling in her purse. She's seen me, and Byron, through the first, bad part. And it must have cost her something. She never looked at Mom's coffin. Maybe she couldn't.

The others surround us. I'm very bad with names. Friends of Mom, and older people, friends of Grandmother. Neighbors from Brooklyn Heights: arty types, banker types. Grandmother's entire Monday Evening Club, moving in under the direction of Mr. Carlisle Kirby.

"Sorry," the men say, like they bumped you in the subway. And handshakes. Quickies, lingerers, grippers, wet fish, double-handed ones that put your hand and your elbow in a vise. Mr. Carlisle Kirby's knuckle-crusher to show what he has left at the age of eighty. And the women who want to say something appropriate about Mom, and something careful. All Brooklyn Heights knows how she died. "You can keep anything here but a secret," Mom used to say. And what Brooklyn Heights knows spreads over the Bridge, infiltrates Manhattan, Westchester, Bergen County, finally ending up in the mental files of people who don't even know Mom first-hand.

The women cup Byron under the chin, try to pull him against them. They can barely keep from patting him on the head. He's taking it okay, but I want to punch them out. Then the tide ebbs, and I can begin to breathe again. The Monday Evening Club moves across toward Grand-mother. I catch a glimpse of Mrs. Schermerhorn's hand touching Grandmother's arm to alert her. Grandmother's head comes up, ready to handle their sympathy. Then she's lost in a small sea of gray heads and lace handker-chiefs.

My best friend from school, Kit Klein, comes in then. His mother's with him. To keep my mind occupied, I try to figure the logistics of this. Did she come to make sure *he'd* come? Or did she tag along with him? I decide he's come on his own. And she's come not because she's pushy or morbid. They live over on Park Avenue, so it's no big pilgrimage. I begin to think about handling sympathy. Just take it, don't ask questions. Don't dissect.

"Listen, man, I'm really sorry," Kit says. He's wearing the school tie like I am, the only ties we own. Little wine-colored shields on a green silk background with gold daggers in between. They're symbols, but of what nobody knows. We're both wearing winter tweed sports coats. Kit's forehead is greasy. He's bigger than me, taller, wider across the shoulders, thicker in the neck. Wrestling team. "This is my mother," he says, remembering not to jerk a thumb at her.

My head goes a little haywire, and I nearly say, "And this is *my* mother." I even start to glance back at the coffin but catch myself.

"We're very sorry," Mrs. Klein says, shaking my hand. I introduce her to Byron, and she shakes his hand too, keeping her distance, no head patting. Then she steps back and melts into the crowd.

Byron gazes up at Kit and me. Whenever any of my friends are around, he thinks we own the world. Kit's stubbing his size twelve into the carpet, looking down into the pile for a clue to deal with this situation. He's a good friend, but we're not too verbal with each other. And he's working hard these days on self-improvement, which is a pretty solitary project.

[20]

"Working on my couth" is his phrase to explain away no longer calling girls chicks, owning a nail file, avoiding his all-time favorite two words: "tough shitsky," buying shirts with thirty-six-inch sleeve lengths to match his thirty-six-inch arms, going from a C-minus with warnings to a B-plus in English. He's starting to think seriously about Swarthmore, Wesleyan, and Williams, in that order, for year after next.

And this is the guy the entire school locker room gave a gift-wrapped can of Right Guard to as recently as ninth grade.

He thinks of something to say. "Your dad coming?"

From Kit this is okay, even with Byron there, drinking in every word. Kit's parents are divorced too. Half the guys at Van Cortlandt are from fractured families. No need for fantasy fathers to keep up your self-image.

"Yeah, tomorrow—maybe. Who knows? Grandmother had Mr. Kirby call him. I guess he said he'd try to make it. But after eight years who counts the minutes?"

Kit absorbs this, identifying probably. But it's really a message I've aimed at Byron in case he's counting on Dad to come. And in case Dad doesn't come, which I expect. Maybe it'd be better if he didn't show. I don't want Byron hassled.

"I could come," Kit says, sort of booming it out too loud, "to the funeral in the morning. If you want me to. I can get out of school. I only got—have history and Spanish in the morning."

"It's Advanced Placement history," I say. "You're just hanging on by your claws. Miss one session and you're down the tube."

"Come on, man." We're both grinning now, and then suddenly remembering where we are.

"No, don't come," I say. "I'll see you Monday." I glance down at Byron. "Or whenever I'm back."

"Okay," Kit says. "I wasn't using this like—for an excuse to get out of school or anything. You know."

"I know. We're big boys now."

He stuffs his hands in his pockets and scratches his legs, making coins and keys jingle. An old habit, dying hard.

"I like your mother," I say. "She's . . ."

"Couth," Kit says.

They leave then. Mrs. Klein signs the guest book. The ballpoint disappears completely into Kit's big paw when he signs after her. There's a little silk tassel on the end of the pen, and it hangs out over the back of his hand while he writes. I can see all this from way across the room.

The thing is, I can't see a few hours ahead to the next day. I can't see my dad.

[22]

THREE

It's a long night in between. Grandmother's Monday Evening Club in full force comes back to the house with us. Almah's on duty. The silver urn's full of Sanka, and there's a long tray of toast points covered with melted cheese and parsley.

When Byron and I go up to bed, Nub's coiled on the landing with the tip of his tail sticking up between his paws. His yellow eyes are at half-mast, seeing and asleep. Byron heads off down the hall to his room. I rack my brain for something to say. "Good night" sounds too formal, too heavy. "You want me to help you get your tie off?"

"No, I can do it."

I stand in my room, slumped against the closed door, thinking about not falling apart in all this quiet. My room. Taller than it is wide, a hundred years old. Older than Mr. Carlisle Kirby, older than anybody in Brooklyn Heights. It still has the trailing-vines-and-morning-glories wallpaper somebody put up in World War I. The morning-glory purple and the vine green are merging into one uniform khaki color. What was this room set up for? Guests. Nobody permanent. I never even thought of it as my room. It's the room I got when Mom and Lorraine and Byron and I came to live with Grandmother after Dad took a walk. It's weird that none of us thought of doing over a room to make it our own.

We were just passing through until . . . Dad came back, or we got a place of our own, or I don't know what. But Lorraine finished growing up in one of these rooms, majored in Ed. Psych. at Hunter for a few terms, and married an Air Force sergeant she met at the counter at Chock full o'Nuts. And Mom lived the rest of her life in a guest room. Two guests down, and two to go.

I think about lying across the bed without undressing, avoid this like a trap, then do it. The minute I'm down on the bed, all I can think about is Mom, like I'm her sole survivor. Two nights ago she was sitting strapped in the front seat of the Buick, undiscovered. Last night she was on a marble slab, being drained. Tonight she's under a lid on Madison Avenue. Tomorrow night she'll be ashes. Even the thing—the disease that was growing in her—

will be ashes. I can cry now in total privacy. So let it happen. Nothing does.

I roll off the far side of the bed and head for the bathroom, dry-mouthed, dry-eyed. Strip off my clothes and put them on hangers on the back of the door because these are tomorrow's uniform too. I arrange the school tie around the crook of the hanger.

The shower's an even more private place. The big white square tiles are veined with millions of hairline surface cracks, the grout dried up into crumbling gray beads. The big steel shower head at the end of the gooseneck pipe is a mile high so you always have to wash your hair when you shower. I stand in there drowning while the water runs down between my legs. The tub fills knee-deep because the drainpipe's slow. The water's rusty-pipe brown, darkening as it deepens. It's like standing up to your kneecaps in hot tea. Then the breeze from the open window starts plastering the shower curtain against my side, and I can't get away from it. So I wrench the knobs shut, jerk the curtain aside, and step back into the world.

I don't plan on sleeping, but I'm asleep almost before I can crawl into bed, still damp inside my pajamas. Usually I don't wear anything to bed. But this is a special night. It calls for all kinds of respectful respectability and caution and protective covering. I sleep for four hours without moving my arm off my forehead.

And I dream. I'm back at Loring and Sons in the room

with the coffin between the floor lamps. It's even worse than reality because this is a dream. The room's completely empty. Even Byron's not there. The guest register's lying open on the table by the door, completely filled up with signatures. The ballpoint with the tassel is out of ink. Everybody who's come to pay respects has paid them and left. Everybody's paid but me. In the dream it's exactly the hour I'm dreaming it—two in the morning. And I've been left behind and locked in the House of the Dead, and it's accurate down to details. Except for the walls. Instead of Loring's dull gold paint, there's a paper patterned with faded vines and morning glories.

I step farther into the room. The lamps at either end of the coffin flicker like real torches. With every move I make, the room gets smaller. My feet sink into the carpet, but with each step there's less floor. The ceiling's coming in too. It's powered by some kind of hydraulic mechanism that lowers it in programmed stages. Edgar Allan Poe computerized. Finally I'm stooping in this room, which is now no wider than a hallway and closing in. There's no place to go except toward the coffin. The shadows urge me toward it. In dreams there's only one direction.

Then we have the first of two big surprises. The coffin lid is open. So okay. I haven't had the complete course. My dues aren't paid because I haven't looked at my mother's dead body. Let's get this dream on the road, and then let's get it over with.

Still, I can't quicken my pace, hurry anything. My timing's off. I get slowly closer to the coffin, which is yawning open. It's all buttoned-down white leather on the underside of the lid, like car upholstery. And now I'm close enough to put both hands on the edge of the coffin. I do. I look up once at the lowering ceiling, and then I look down into the coffin.

And here's surprise number two. It's not Mom lying in there at all. It's somebody else. There's been a vast cosmic mistake. The entire dream changes mood. It's not even terrifying. It's too confusing.

There's a man lying there in Mom's coffin. I know he's dead. People often die with their eyes open. He's wearing a sports shirt. Ivory palm trees on a red background, open at the neck. Short sleeves, hairy arms, a big guy, completely filling up this woman-size coffin. I can see the rim of his teeth through slightly parted lips. This is not Loring and Sons' best work. He needs a shave, but he has a hell of a good tan.

And I can't just walk away and try another floor at Loring's. Because this particular corpse is my dad.

The knob on my bedroom door turns, and I'm awake and half out of bed. One heel hits the floor. I fight off the sheet. The digital clock says two-oh-seven and flips to eight. "Who is it?"

"Jim?"

"Nub's not in here, Byron."

"I know. He's still on the landing. . . . Can I come in?"

"Sure." The door opens wider, and there's a trapezoid of dim light on the floor. Byron's shadow fills part of it, elongated. Still, he doesn't come all the way in. I'm completely awake, glad to be out of the dream but I'm not sure what I've got here to deal with. "You want the light on?"

"I guess."

He's standing in the doorway, small hand still clenched over the knob. He's got this destroyed look on his face. This is natural, I tell myself. He's been too poker-faced up till now. I stare into this little kid's old face, looking for a sign. Then I see he's wearing pajama tops and a bathrobe and no pajama bottoms. Little white spindly legs disappear down into high-sided bedroom slippers. We stare at each other.

Finally he says, "I couldn't help it." His voice cracks in every direction. He's too upset to cry. Nobody at the age of eight should be too upset to cry. His chin keeps bucking back against his throat.

"You want to vomit?" I say, standing up.

"I want to, but I won't." And then it's breaking over me. He's embarrassed. I've seen this in him before, but not often, and I wasn't looking for it now.

"What happened? Come on over and sit down on the bed."

He pulls back, almost hides behind the door. "No.

[28]

I—I woke up. And . . . I'd wet the bed." His face goes all out of shape. He's shamed himself and dirtied himself, and it's too much for him to handle. He doesn't see that it's part of the bigger thing—reaction. In a way it's almost a relief, at least to me. I walk over to him, and he cowers. I can't help it. I'm standing over him, and I'm enormous to him.

He's shaking his head and looking up, and I see the whites of his eyes. "Why did I do it? I haven't done it since I was little. So why?"

"Look, it doesn't matter. Nothing that happens tonight or tomorrow counts. Just take my word for it." I've got my hand on his shoulder. It's like a bird's shoulder, full of webbed bones. "Listen to me, it doesn't count." I give him a shake for emphasis. "Come on, we'll change the bed."

"I don't know where the clean sheets are," he moans, and every word falls apart. If this kid could break down and cry, it'd be like a summer rain. I turn him around, and we walk down the hall to his room. Past Mom's room. Past Grandmother's. We won't be disturbing her sleep. I know that she's in there lying fully conscious with the unopened Valium bottle on her bedside table and her cane under the bed, at the ready, ready for morning. I know she hears us. And she knows I'm coping with Byron, while she's coping with herself.

He stops at the door of his room, suddenly doesn't want me in there. I reach over his head and run my hand

[29]

down the wall for the switch. The overhead light clicks on. It's a regulation room for this particular house. Blistered wallpaper with some secret-code design, and an odd-lot selection of furniture featuring a brass bed. There's a sink in the room with a tacky square of linoleum under it. Some ancient homeowner's idea of a handy guest-room facility. Byron's pajama pants are wadded in the sink.

He's pulled the top sheet down off his bed. There's a big stain in the middle, already soaked in. No smell yet. I walk around him, very brisk. Big brother taking over. "Pull the sheet out at the bottom on your side," I say. We fold the sheet back, flip his pillow off on the floor. The same stain's on the quilted mattress cover underneath.

"Oh, no," Byron whispers.

"That's what mattress covers are for." I hope, when we get it off, the stain hasn't gone any deeper. "Reach under the mattress over there. The cover's attached with elastic across the corner." He heaves up the mattress, and the cover curls up. We fold it, and underneath on the flowered mattress there's a barely damp spot the size of a Frisbee. "That much won't matter," I say. "Get a wet washcloth."

I scrub away at the mattress, trying not to make a bigger wet place because Byron's fixated on it. "This'll dry," I say. "No harm done."

We find clean sheets out in the hall cupboard but no

mattress cover. I lay a towel over the damp spot on the mattress and we make up the bed. I start to jam the sheets in on my side, but I notice Byron's squatting and making hospital corners on his side, the way Almah makes up our beds. So I pull my side free again, fold the corners into proper patterns. When we finish, the mattress looks like a gift-wrapped box.

I fold the stained sheets in on themselves, trying to make a bundle that's dry on the outside. "What'll we do with them?" Byron asks. "I don't want Almah to know."

"Forget about Almah," I say.

"She'll *find out*," he says. "She won't like it." But we know that's not what he means. He doesn't want her to know or anybody to know that he's suddenly reverted to babyhood. He's eight years old, as old as he's ever been, and every minute of it should count in his favor. Now it's all erased, and he's back at the beginning. I know exactly how it must feel, but I can't say so.

He hesitates by his bed directly under the hard light. If I tell him to, he'll crawl in and try to stay awake the rest of the night so he won't shame himself again. "Come on down to my room. Bring your pajama pants." I carry the bundle of sheets down the hall, and he follows. I tell him to take a shower in my bathroom. Then I start washing his pajama pants in my sink.

He gets into the shower and pulls the curtain across. Then he stuffs the robe and then the pajama top out over the edge of the tub, and they slip down on the floor.

He's gotten suddenly modest lately. He showers a long time while I scrub away in the sink, doing a job on his pajamas with a nail brush and Irish Spring soap. I can see his outline through the plastic shower curtain. He's standing at attention under the shower, which is needling down on his head. He never raises his arms. He never shifts position. It's the shape of a little concrete gnome in a rock garden.

When he finally shuts the water off, I drop a towel over the top of the shower-curtain rod. Then I reach around and hand him his pajama top, then his robe. I wait until he pulls the curtain back and steps out of the tub, more or less fully dressed. It's almost funny, but I don't try to play it for laughs.

Instead, I'm standing there anticipating his next problem: what to do with the sheets. They're too messed up and wet to stuff in the hamper. I have a brainstorm and lead him back into my room. Then I hang one sheet out of each window. He watches this as if he's witnessing the invention of the wheel. "They go in the hamper in the morning."

"Almah will know," he says. "She counts everything."

I let this pass. "Get into bed," I say.

"Your bed?" His eyes are round. At rare moments his eyebrows almost meet in the middle. Now they're doing this.

"Sure."

"What if I—"

"You won't." I flip out the light.

[32]

It's not quite as wide as a double bed. We both shift around, trying to give the other as much room as there is. "I don't need a pillow," he says, pushing it my way. He needs a pillow more than anybody else I've ever known. He sleeps completely curled around one. I let him give me the pillow.

"Almah," he begins again, "she—"

"Byron, stop a minute and think, okay?" He lies there, suddenly mute. "Think about Almah if you want to. I mean really think about her. She's at home right now, down in Red Hook, right?"

"I guess so."

"She's worked for Grandmother for—I don't know how long. The beginning of time, right?"

"I guess so."

"And if Almah ever liked anybody in this whole family, it was Mom. Right?"

Silence. Then, "Yeah. She never said, but I think she liked Mom pretty much. When Mom was real sick, she'd stay late and do extra things."

"So how do you think Almah feels tonight?"

Byron lies there like a little mummy, wrapped in his robe, nose straight up toward the ceiling. He's giving this his complete consideration. "Bad?" he says finally.

"Really bad," I say. "Because she's the kind of person who can't show us how she feels, apart from complaining. I bet she's in her room down there in Red Hook staring at the walls."

There's a long silence then. If it was anybody but

Byron, I'd think he'd gone to sleep. But then he says, "I know what you're saying. Almah's too upset right now not to be one of us, whether she wants to be or not."

"That's it."

"She might even be too upset to count the sheets," he says in a voice slipping toward sleep. That's all either of us knows till morning. When I wake up, Byron has my pillow, and it's tucked into his chin, and he's sleeping furiously into it.

FOUR

Fifteen minutes before the funeral, and something's already gone wrong. Mrs. Schermerhorn and Mr. Carlisle Kirby are fanning out from Grandmother. Mom's coffin has been moved. It's down in front of the pews. The white rosebuds on the lid are upstaged by all the "floral tributes" Grandmother expressly asked people not to send. Big salmon-pink spikes of gladioli, yellow, heavy-headed chrysanthemums, other things I can't name. There's even been a mutiny in Grandmother's own ranks because the Monday Evening Club's sent a spray of red, white, and blue carnations.

People are already beginning to come in, settling behind some invisible line along the rear pews. I'm fairly confident that my dad isn't going to show. He'd surely be here by now. I think I'd recognize him if he was here. Haven't I just seen him a few hours ago wearing a sports shirt in Mom's coffin?

Mr. Kirby is having words with the undertaker, then three undertakers. His long, blue-veined hands are working in the air. His crooked finger's going to start tapping an undertaker's vest button any minute now. Artificial stained-glass daylight is striking sparks off his old bald head. The white fuzz growing out of his ears is quivering. He looks like an Old Testament prophet in a Brooks Brothers suit. And the undertakers are listening to him.

The problem is they've slotted us into exactly the chapel Grandmother didn't want. And she knows them all. There's no side room or curtained alcove for the family to sit in during the service. We're expected to sit down front in full view.

"Do I need to remind you people this is a *funeral*, not a wedding?" Mr. Kirby's outraged whisper carries all over the chapel. The undertakers are trying to hold their ground. "Do you people actually expect Mrs. Livingston and her family to sit out front, exposed to public scrutiny?" His whisper drops lower and gets more vicious. "Do you people *realize* how this looks?"

This could be bad for Loring and Sons' business. This is what Mr. Kirby means. His arm sweeps over the Monday Evening Club now taking up a complete pew, end to end. He doesn't actually have to spell out that they're all candidates for Loring and Sons' services in the near future.

The undertakers are computing this, letting Mr. Kirby have his say because they have no choice and because they haven't figured a way out of this yet. They aren't used to dealing with old men. They're used to burying old men. It occurs to them that Mrs. Livingston, Grandmother, has her own linebacker. He looks like Don Quixote, but he's got the impulses of a Wall Street lawyer who never admitted to retirement. He keeps going for the jugular.

But this is rush hour at Loring's. Every chapel is going full blast. A compromise is worked out. Screens are brought in on rollers, like the screens around Mom's bed when she was in the hospital ten, fifteen times. Mom still hasn't made her escape from all that yet—another hour to go. Even the flowers smell like a hospital room.

The screens are overlapped to make a little roofless room for us down front. This slows the countdown. Clocks stop. It's ten minutes after the hour before the organ begins to play, automatically, "Evening Star" from *Tannhäuser*.

Grandmother's not going to put herself on display. That's out. And she's trying to screen out what she can't foresee. She hasn't had a word with me about Mom's death. And she's kept even farther from Byron. How can she know how her grandsons will react? What if we disgrace her? She's depending on me to keep that from happening. But she likes a second line of defense. Is she worried about breaking down herself? No.

We're all rounded up and led down to the screens by undertakers moving like this was their own idea. They've left two empty folding chairs beside Grandmother, and they shepherd Byron and me toward them. But Grandmother pulls Mrs. Schermerhorn down on one side of her and motions Mr. Kirby into the seat on her other side. She's covered her flanks. Byron and I move into the row behind.

Grandmother's Dr. Painter is assigned an end chair in the enclosure. She'll show him what he can do with his Valium. One of Mom's doctors is there too. Grandmother knows how to show people their own limitations.

One of her distant cousins is in the doctors' row behind us, given a place because she's ridden the bus all the way down from Utica and deserves something. I don't know her name.

The service begins, conducted by a Congregational minister probably picked out of the Yellow Pages. Grandmother wouldn't have her own Episcopalian priest officiating. Would this be putting him on the spot because Mom was a suicide? I don't want to know the answer. Theology is not my thing. Byron sits silent and upright. His necktie falls in folds halfway to his knees over his clenched legs.

[38]

The sermon's in third person, with long Biblical quotations that bypass death and linger over the geography of the Holy Land. Sheep grazing on green hillsides above domed towns.

It's not hard to take, so I let my guard down and conduct my own service without benefit of clergy. She was dying anyway. A long, no-win, no-known-cure sentence of death. They couldn't even give her a set number of months. That's only on TV—daytime TV at that.

I found out Mom was dying by seeing it in Grandmother's face. Firm lines going hard. Mom worked at not showing it. She used up too much of her reserve building a brave face when she came home from the hospital and the pointless treatments, every time with more makeup, finally almost a mask. Toward the end she was moving slow. It must have taken her a long time to get out to the garage.

Actually she's been dead a long time. This strikes me as a big revelation. Either that or I'm trying to put distance between me and the event.

Her talk always turned away from herself. She'd ask me more about what was going on at school than I knew, sidestepping anything that sounded like Last Words. She read to Byron a lot, and he let her. *Charlotte's Web* more than once, when on his own time he was reading *Watership Down*.

When she got worse, I knew it mainly by seeing the light from her room making shadows down on the back

[39]

garden and across the garage wall. Later and later into the nights. She was Grandmother's true daughter—hanging tough. Toughest with herself.

So she died when she couldn't pretend to live any more. She took her life in her own hands—the real meaning of that saying. Let's have no talk about heroism and cowardice here. This is the way it was. Why didn't she leave us a note? For the same reason she didn't lay Last Words on us. Why didn't she just take a handful of pills and die in her bed so it could look natural? She was way past pills, that's why. They couldn't touch her. She'd taken too many. Her body was a drugstore that had lost its lease. Does this cover the ground, tie up the loose ends? No. But this is as far as I go. I'm not ticketed to the end of the line.

Finally I work around to why Grandmother has gone out of her way to find this anonymous minister. I'm a plodding thinker, but I get there. He'll preserve our privacy better than screens. He won't look for answers to questions we don't want to ask. He's a sure thing.

But he's not finished. We're not quite off the hook yet. Why have funerals when they only prolong the agony? His voice drops, changing the rhythm. He's closing the Bible. It's time for the prayer, and we're on the home stretch. Mrs. Schermerhorn reaches down for her purse. Grandmother's head in front of me begins to bow. But it seems there's going to be a slight variation. He's not ending with a prayer. He announces that he's concluding with a passage from Milton: *Samson Agonistes.*

Better yet—not even Biblical at first hand. He quotes from memory in a low-pitched voice, drawing us in:

"Nothing is here for tears, nothing to wail
Or knock the breast; no weakness, no contempt,
Dispraise, or blame; nothing but well and fair,
And what may quiet us in a death so noble."

Only four lines, and he's kept his voice bland, no stage business. Mrs. Schermerhorn's hand remains unmoving within reach of her purse. Is this a prayer, or isn't it? Her hand seems unsure, even through the glove. Grandmother's head is still bowed.

Nothing is here for tears. But they're there anyway, setting little fires under my eyelids. *No weakness, no contempt, Dispraise, or blame.* No, no blame. She was dying anyway.

A death so noble. What are we supposed to do with this? Noble—to leave Byron and me? He's only a kid. Every day matters with him. He'd already been left by one parent. One he doesn't remember and now one he does—two quick jabs to the groin, both times when he wasn't looking.

No, that's the wrong direction; take another. Noble—not to let us see her turn into a vegetable shriveled with pain. Is this better? No, it goes nowhere. I can't buy it. Put that last line together and run it by again. *And what*

[41]

may quiet us in a death so noble. I've got a handle on it now. The minister can call Mom's death noble because he doesn't even know how it happened. He's the only stranger in the room. But he's the one at the microphone. Later the rest will find their voices: poor Mom, poor Grandmother, poor sons, poor daughter way off in Germany, a shame, *the* shame. But right now we're all being purified by poetry, and silenced. So Grandmother knew what she was doing.

It's over, with no filing past the coffin, no pallbearers. Feet shuffle outside the screen. Grandmother waits until these outsiders realize we're not coming out to be sympathized with again. Then Mr. Kirby stoops for Grandmother's cane, puts it in her hand. We stand up: the two of them, Mrs. Schermerhorn, behind us the two doctors and the cousin from Utica, Byron and I. The eight of us.

No. Nine. Somebody's slipped in late, next to the cousin. A stocky guy, my height. Blondish hair going a little white at the temples. Rumpled Palm Beach suit, narrow lapels, narrow tie, a shirt collar point caught on the outside of the lapel. Probably a one-suit man. Winging out from his eyes are deep creases paler than his tan.

Grandmother pivots on her cane, ready to leave. She catches sight of him, and it throws her timing off, no more than a second. Mrs. Schermerhorn's eyes dart back and forth.

"Well, Howard," Grandmother says to the stranger, "these are your sons, in case you don't know them."

Riding backwards on the jump seat in Mrs. Schermerhorn's Cadillac, I watch the grille of the Lincoln limo from Loring's tailing us. Across town, around the lethal approach to the F.D.R. Drive, gaining a little measured speed down past Sutton Place and through the U.N. tunnel. Past the helicopter pad and then curving out toward the East River between Stuyvesant Town and the tennis courts, under the graffiti on the pedestrian overpass. The tugboats are plugging along on our level, curling the foam with their bows.

I watch the grinning grille on the Lincoln between Grandmother's face and Mrs. Schermerhorn's. Byron's glued to it too. The limo windshield's a mirror, beginning to reflect the Wall Street skyline. Behind it is the Loring driver and behind him Dad. Nothing real about this. It's only—traffic.

In our car absolute silence. Mrs. Schermerhorn's there because she knows not to fill up awkward silence with awkward chitchat. With Grandmother in the back seat, it's easy to forget this is Mrs. Schermerhorn's car. Grandmother's face is set. But upon what I don't know. Her plans haven't gone completely out the window. Right now we're into her alternate maneuver. She hasn't counted on Dad showing, but since he has, she can cover this too. No curbside rejections. No scene of any kind.

[43]

He's to come back to the house for the funeral lunch, break bread with the sons he dumped. Squirm under the pitiless eye of his ex-mother-in-law. Pay.

The Fleetwood's tires hit the metal mesh on the Brooklyn Bridge. The bracing cables of the bridge flash past and ripple the Lincoln's hood behind us. We're still being tailed. Lose them, Nathan, step on it. Let's see what this old Fleetwood can do if you open her out for once.

Nathan brakes gently, and we take the first exit off the bridge, spiraling down toward Cadman Plaza. We swing into Montague Street, double back toward the river. Nathan hauls on the wheel. The Fleetwood steers like a semi. Two more sharp lefts, and we're home. We take up both parking spaces in front of the house. The Lincoln idles in the traffic lane, blinkers flashing. Nathan hands Grandmother and Mrs. Schermerhorn out. They wobble on the sidewalk, find their land legs. The sunlight turns the Lucite cane into a fiery rod. Before I can unfold my legs and get out of the car, Byron's hand is bunching my coat sleeve.

"Is that really Dad?" he whispers. "Ours?"

Almah's sister comes in to help serve the lunch. Rustling black shiny uniforms, Earth shoes, white aprons, no frills. Almah's face is a sculpture, carved with primitive tools. She has a jaw whittled out of some incredibly hard wood. Eyes that can spot trouble at tremendous range. Her sister tries to keep behind her at all times.

[44]

We don't fill up all the space, even when the whole Monday Evening Club reassembles. The average age now gathered together comes in at about seventy years, even figuring Byron and me in. And Dad. There's a knot of people in the front room, between the bay window and the Bechstein grand piano. Another couple of groups cluster in the windowless middle room, being stared down on by the portrait of my dead grandfather, looking like J. P. Morgan. The back room, always called "the summer dining room"—year around—looks down to the back of the garage wall across the garden. The buffet table's set up in there, still bare. All the flowers that have been sent to the house are down in the cellar, beginning to die of thirst.

People stand, then sit in isolated clumps with distances between. It's the configuration of a subway car. There's an understanding that Grandmother isn't going to play hostess. She sits in the wing chair under Grandfather. Almah backs through the kitchen door with a tray of pre-poured sherry and two tumblers of milk. One milk is for Byron, the other's for Mr. Kirby's ulcer. It's having a workout. He's on his feet but wavering, tall and hunched, in the middle of the middle room, halfway between Grandmother and Dad, who's barely inside the hall doorway.

I start walking through a lot of manufactured conversation toward Dad. He's not drinking the sherry. He's looking around the room, but he knows I'm getting closer. I remember Winifred Highsmith: "I never can keep track of people's children." "I'm Jim," I say to him, to Dad.

"Yes, I know. I had time to—work that out. I was sitting behind you. At the funeral." I wonder if our voices sound alike. You can never really hear your own voice, not even on tape. You're too occupied in thinking it isn't an accurate reproduction. I decide our voices are nothing alike. And his hair's a couple of shades lighter than mine, except it's got some white in it. We're not even built the same. Except I'm not forty-six.

"I'm seventeen." Should I say "*sir*"? Too stiff, almost a sneer. Nothing's right, but I didn't walk across this room to freeze him out. Why did I walk across this room?

"Seventeen last February," Dad says. It's eerie, annoying. Like having your palm read by a stranger who's a lucky guesser. "You were born in Barnes Hospital, during a snowstorm. The only blizzard they'd had in St. Louis for years. We drove in as far as Clayton without snow tires, then got a cab somehow the rest of the way. But you took your time. You weren't born till the next day."

This is starting a ridiculously long way back to lay the foundations. But safe. "I didn't know that," I say. "I knew I was born in St. Louis. I have to put that down on forms for school. But I didn't know about the blizzard." We stand there looking at each other. I think I can read his mind. Shall we go on with this, recounting where and how Lorraine and Byron were born? We could fill up a lot of time this way. We were all three born in different places while Dad was moving around, climbing the corporate ladder of some company. That was before he fell off the ladder, or jumped. I see by his face that calling the roll of births isn't the way to go.

It occurs to me that the reason there's nothing familiar about him is that (a) I was nine when he left and so he looks shorter now, and (b) I've never seen a picture of him. He got erased from the family.

"Lorraine's in Germany," I say. "She's married."

"I know. There are always people who tell you things." He doesn't say who these people are, how the network for information works. It sounds complicated, spotty. "What are you thinking?" he says, catching me well off base.

"I don't know," I say. "I'm just taking every word as it comes. They're not exactly flowing."

"I was thinking about the first time I was ever in this room," he says. His eyes move around, skipping over Grandmother. "I followed your mother home one day. Nine hundred miles. We were both still in college. I met her at this resort in Saugatuck, Michigan. I washed dishes—ran the dishwashing machine. Barbara—your mother—worked in registration."

The conversation keeps spiraling back in time. He'll be back to his own birth at this rate.

"I think I must have been just about as uncomfortable

that first time here as I am now. Some things never change. You never really grow up."

This line is too easy. I let it pass. It's creeping up on me that this guy has not turned up after an eight-year coffee break ready with excuses and a smoke screen of big talk. I was expecting something a little more overbearing. His eyes still roam around the rooms. There's a pointlessly big chandelier in the middle of the ceiling. A denim bag was put around it in preparation for summer. But Almah took the bag off early in the morning, and Nathan came over to help. Dad looks at it, and then his gaze drops to a point behind my shoulder, where Grandmother's sitting.

"Still a lot of starch in her upper lip," he says.

Almah and her sister are bringing the food into the summer dining room. Almah's carrying a long tray of chicken-salad sandwiches—it's always chicken salad—and the tray isn't buckling under the weight. Her sister's struggling with the Sanka urn. I think about rounding up Byron. But I don't see him. Mr. Kirby's milk glass is still in his hand. The other glass, drained, is on the piano.

"I've got to go find my brother," I say, ready to walk around Dad. Where's Byron? I don't want him left alone. I let him out of my sight for one minute, and he—

"Can I help you look?" Dad says.

"No." I'm at the foot of the stairs. And I wait long enough to keep Dad from falling in behind me. Then I head up.

Byron's in his room, hanging around by the front window, but not looking out. The light's tricky in there, north light bouncing off the brownstone fronts across Remsen Street. It's a moody, late-afternoon kind of room. He's got one hand on the sink. I wonder if he's been sick. The bed we made up so neatly in the middle of the night is still tight and untouched.

"What are you doing up here?"

He looks up, and the light halos all around his slicked-down hair. "I didn't know where else to *be*," he says in a strangled voice. "So I just came up here. Where was I supposed to *be*? You didn't want me talking to *him*, did you?"

Probably I didn't. Big brother runs interference. Big brother moves out first from base camp, checking unknown terrain. Big brother checks swamp for deadly quicksand. "Come on," I say. "It's time for lunch. Get washed up. Roll up your sleeves. You didn't eat much breakfast. You're probably hungry."

Byron's eyes flick away from me, to the door. Dad's standing in it. I don't know what I feel about this. Not a regular kind of anger.

"Is he all right?" Dad says, practically whispering. He doesn't know this little kid. He hasn't made the contact. I'm standing there between them. I point at the sink, and Byron reaches for the soap. Don't try to walk around me to get to him. But Dad doesn't try. "Can I see your room?" he says.

"What for?"

"I'm in no hurry for the chicken salad," he says. "I never was."

I lead him away from Byron back through the length of the house, open my door, and let him walk in. "My God," he says, "the same wallpaper." He scans all the morning glories and vines. I wonder what he was expecting. Pennants from all the Ivy League schools? Punk rock posters? Complete collection of vintage beer cans?

"This was the room I stayed in that first time," he says. "I had the distinct impression that Grace—your grandmother—sat out in the hall all night with a musket across her knees."

One more cheap shot at Grandmother, and I'm either going to come to her defense or defect to his side. Why this is turning out to be the big moral issue of the day amazes me.

But Dad walks across to one of the rear windows, looks down at the garden and the tar roof of the garage. He stands there with his hands locked behind him, wrinkling up his coat. Byron stands that way sometimes, like a little old man.

We can't keep things on this level, I think. Not through lunch. It's a no-man's-land, nothing clear-cut. And nobody's going to take him off my hands. I move up behind him. He's still looking down through the dogwood branches in full flower.

[50]

"That's where she died," I say. "Mom."

"In the garden?" He turns on me, this astonished look on his face.

"No. In the garage."

"In the *garage*? What are you talking about?" His hand comes up to grip my arm. But we haven't touched, and don't. There's a wall there. "How in the hell did that happen? Didn't she die of . . . Wasn't she in the hospital when . . ." His voice trails off, but he's waiting for an explanation. And I see now he needs one. It was Mr. Kirby who called Florida to tell him Mom died. That's all he told him.

"Mom killed herself. In her car."

And then Dad's face goes all out of shape. He has this destroyed look. Byron gets it sometimes.

FIVE

Monday morning. Dad's gone back to darkest Florida, leaving no trace. And no Final Words; at least he had one thing in common with Mom. I come down dressed for school in case I'm going.

I'm going. It's written all over the dining-room table. The table groans with routine, business-as-usual. Grandmother's not downstairs; she never is for breakfast. Byron gets up an hour later since he only has to walk around two corners to get to his school, Brooklyn Heights Collegiate. We're both going to school. Two hot-cereal bowls are set out. Oatmeal till the last day of school even if there's a killer heat wave, even if there's a recession in rolled oats. Almah's making stirring noises in the kitchen.

There's an airmail letter from Lorraine—weepy, suffering. I can't get through it, set it aside, turn to the next item. The application for a summer camp counseling job I mislaid somewhere has made a comeback by my juice glass. Grandmother's hand at work the night before. What do we have here? Name, age, previous camp experience, major sports, sports qualified to coach, waterfront skills, other talents please list for example storytelling, auto mechanics, dramatics, crafts, musical instruments, woodworking, specify power-tool proficiency. References. Five dotted lines for biographical sketch.

Tell us your hopes, your heartbreaks, your nastiest vices, and the college of your choice, and maybe we'll pay you a hundred a month to teach twelve-year-olds beadwork when they'd rather be drag racing.

There's also a pile of literature beside Byron's place. I read brochures upside down, topped with *A Parents' Guide to Outdoor Summer Experience in the Tri-State Area*. Separate folders for camps in Connecticut, Upstate, the Ramapos. We're going to get through the next few weeks somehow, and then we're going to be packed off. The summer's as far ahead as even Grandmother can see. I don't see much past it myself. Another year and four months, make that sixteen months, and I'll be ready for college. But I can't see that far, and where does that leave Byron?

But back to the present. Have I got a prayer of landing a camp job this late, even Camp Candlewood in Boondock, Vermont? Yes. Mr. Carlisle Kirby is on the camp board. I'm a shoo-in. The place on the application for references has been x-ed out for my convenience. The Gray Panthers strike again. The elderly can't deal with you, but they sure as hell can deal in your behalf.

Even as I shuffle through the application form, Mr. Kirby is gearing up for Mom's inquest, drawing in Grandmother's lawyer so he won't feel out of it. He puts his own lawyer on call, ready to coach Nathan if his evidence is challenged or if Authority leans on him. The Family is to be spared all known hassles.

I don't know all this at the breakfast table. It leaks out in dribbles over the next month, even after the inquest is a . . . dead issue. You hear things in an old house. The walls talk. Stairwells echo. Even the closed door of Mom's room mutters. A word here and there wafts up from the middle parlor when the Monday Evening Club's in session. Mr. Kirby uses our phone, which, weirdly enough, is built into an ancient phone booth under the stairs in the hall. Mr. Kirby speaks in a carrying voice because he's slightly deaf and doesn't know it.

I learn nothing from Grandmother. Zero. Even if I wanted to, there isn't enough curiosity in the world to breach her defense. Inquest, coroner, suicide, even death itself are not in her working vocabulary. And she wasn't put on this earth to be interrogated by a teenager. Or anybody else. Byron seems to have been born knowing this, and more, about her. But I have to keep relearning. On the subject of Grandmother, I'm remedial.

That first school morning I think her walls will never crumble. They will, but I don't know this yet. Life sends you damned few advance warnings.

Almah brings in the oatmeal, ladles it from a great height into the bowl. No raisins involved in these plops, not at a dollar twenty-nine a pound. We run a tight ship here. I hear Almah thinking. This is the fifth morning she hasn't had a breakfast tray to make up for Mom. There's still a hole gaping in Almah's day.

The subway stinks worse than usual. I'm in the front car where the same six girls are already on, talking around each other down the long seat. Why don't they ever talk to the ones nearest them? Their hair whips as they crane out past each other, firing conversation down the row. They go to some school that doesn't require uniforms, always leaping up in a bunch and exploding out the door at 72nd Street and ganging across to the local.

We've practically grown up together in the front car. But I don't know them. I only know strategic parts. They wear sandals today, and the one they always talk around paints her toenails purple-red. What is this? Compensation for being the group dodo?

All their skirts are too long and getting longer. I've committed their thighs to memory, but memory begins to blur. One girl doesn't shave her legs, almost doesn't need to. Peach fuzz, no bristles. How would Kit Klein in his uncouth days describe these legs? If skirts get any longer, I'll end up ankle-peeking like Mr. Carlisle Kirby in his youth.

We stop dead center in the middle of the East River tunnel. Complete power loss. Dead, underwater silence. The lights dim. Oceangoing vessels pass directly over our heads. Green Mafia corpses, scientifically weighted, graze the tunnel top with seaweed hands. Rats glance out from tile chinks at the sudden stillness of us becalmed underwater.

Then there's a bone-rattling take-off. Zilch to twenty miles an hour. The girls shriek. Notebooks slide from lap to lap. Knapsacks keel over. We're still going to school, won't even be late. They groan, roll their eyes. Even look my way. Small disasters permit this. But I keep their looks, don't even give them back. Inmates of boys' schools are misers by nature, who save such encounters to feed their fantasies in closed-door nights.

School's a blur, already winding slowly down to end-of-the-year. Kit Klein appears down a hallway, head and shoulders above the crowd. Sends out various types of hand signs to me. The general meaning being, you okay? Got through it all right? Great, see you later.

Mrs. Berger in English is especially gentle. We've been doing poetry. She'll give me the mimeographed sheets I missed. Everybody else turned in his paper on Modern-

American-novel-of-year-choice Friday, but of course I can take my time handing mine in.

"Tomorrow," I tell her. She's looking up from her desk at me, through a haze of black hair, silent sympathy sent out in microwaves.

"Take your time," she says again, softly. Her breasts are a major handicap. Big-breasted women who teach in boys' schools have to struggle for eye contact. Some of her students have never seen her face. She has a fantastic wardrobe of high-necked blouses, loose jackets, bows under the chin. Nothing works. "What a pair," Kit Klein still says even in the pursuit of couth. "What a pair . . . va . . . va . . . va . . . VOOOM!"

I pass on from Mrs. Berger to trig, Contemporary Social Issues, chem, gym. Somehow it's Friday, then Monday again. For the first time I actually comprehend that routine isn't all bad. It sees you through. Bells ring and you move. Bells stop ringing and you sit down. The little guys I coach have their own way of handling me. They can't deal with death, so they don't mention it. And then they forget. We convoy across Broadway, into the park. The usual squabbles break out.

They get worse as we near the end of school. After all, we're the Van Cortlandt Academy farm team, perpetually in spring training for a season that never comes. These are not the Boys of Summer. We'll never play Shea. And they don't honestly believe they'll ever be upper schoolers qualifying for the Real Team. I'm their

[57]

only link with this Big Time. Upper schoolers, even me, are Supermen, have the world wired. Able to knock the ball out of the park, out of the Bronx, out of the solar system. So they wait hopelessly and listen restlessly while little Ricky Hatfield, clown of fifth grade, does his Howard Cosell imitation for the millionth time, and then they resume squabbling.

We train less, exercise less, worry about finesse less, and play more ball. When we're up to strength, we can just field two teams. With our own rules. Giants and Dwarfs, by name, which has nothing to do with comparative heights. There are giants and dwarfs on both teams, evenly divided by me. There's no grabbing up a bat handle to choose players: this creates rejects, and I don't want any water boys.

Our own rules are strictly our own. If the Giants get a double and a single and still don't score, while the Dwarfs get only two singles and don't score, the Giants win. And vice versa. No overtime. Mysteriously, the floodlights stop going on, possibly to honor late daylight, possibly to honor Con Ed.

They like our private variations in the rules. It sets them apart, makes them more clubby, and boosts morale until they can get to their various summer camps. Then they can lie like bandits about this great school team they star in.

I come home one night fairly late, stopping off first at Burger King. We don't pretend to have family meals any more. This goes back to Mom's sickest days. Almah gets Byron fed before she goes home, and that's all Grandmother or I care about. I've lost some weight since Mom died, and I don't have any to lose. Besides, the annoying headlines are blaring in my head again. SUICIDE'S SON PINING AT SENSELESS LOSS EXPIRES OF HUNGER. I come home carrying a second chocolate shake, double thick.

The front hall light's off, and I hit the stairs at my usual top speed. I nearly run Grandmother down. I actually bump into her, our first human contact in years, maybe ever. Her cane clatters against the banister. We grab each other. Somehow she keeps her cane, and I keep my chocolate shake. Two other people would get a laugh out of this. All we get is a sharp intake of breath. We're all tangled up. I step down a step, see that she isn't going to topple forward. I can't tell if she'd been heading up or coming down.

Then I begin to see. She's been standing there in the gloom, waiting for me. It's true we have a lot to discuss, now that Mom's no longer there in between us. Like, where do we go from here? After the summer, after we start crawling out of our post-funeral shells into our normal-size larger shells. But redefining relationships with Grandmother? Getting it all together with Granny? I can't see it. It's science fiction. So what are we on the brink of?

I wait while she pulls herself together. The cane shifts

from hand to hand, one grip on it, the other on the
railing. "Jim, I thought you'd better know. Nathan was
good enough to sell it for me."

Her voice is weak and strong at the same time. But
what's she talking about? "He took it someplace out on
Northern Boulevard. And he received seven hundred
dollars for it. I wanted him to take something for . . .
all his trouble, but of course he wouldn't hear of it."
Her voice gathers strength at Nathan's strength of
character, which is already well known. She's giving my
slow boy-brain time to know what she means without
having to spell . . . it . . . out. Northern Boulevard,
in Queens: used car territory. Nathan's sold Mom's
Buick Skylark. I have a quick flash of him easing into the
seat where Mom died and nosing the Death Car up the
Brooklyn-Queens Expressway ramp.

"—a good price for it, I think."

Her cool is half shot if she's talking money. Money is
another of her taboo topics. But she knows I haven't
been near the garage, won't go. And she thinks I better
know what's happening. You can talk about a car, even
this one, with a boy because cars belong to the man's-
world category of points, plugs, buying, selling, trading
off.

She's still speaking, in starts and stops. "—divided
equally between your savings account and Byron's."

This is our entire inheritance, though she doesn't lay
this on me. We've lived off her for years. And all the

hospital bills "threatened her capital," in Mr. Kirby's crisp, money-crackling phrase that floated up the stairwell more than once.

This is it then. She wants me to know the car's sold. Period. Not even any heavy advice about keeping my half of the take untouched in savings. I could draw out my entire three hundred and fifty dollars on the day of deposit and blow it, and she'd never seem to know. Because if she starts trying to raise us, she'll be committed. We're orphans now in her eyes. Dad never counted. Orphans need care, but cars can be sold. I understand this. I'm glad the car's gone in case I ever have to go out to the garage.

She's heading downstairs now, I see. Moving around me with impatient little twitches of her cane, all her hurt nicely bottled up except for that unconscious wrist action on the cane's crook.

We may not find anything to say again for days. So over my shoulder I say, "Nathan got a good price for it."

She observes a moment of silence. Then, "Well, I wouldn't know. I would simply assume."

The next day I get called out of Contemporary Social Issues, last period of the day. There's been a phone message for me. The school office is having a tense moment. Secretaries' eyes fall on me and slew away. Trouble brewing, already brewed. What in the hell have I got left to lose? There's been a call either from Grandmother or Long Island College Hospital. And I'm supposed to go either there or home. We're not sure about this. I'm supposed to do something right away. The head secretary is giving me this garbled message from her shorthand pad. Her shorthand is perfection; it's the message that's giving her trouble. Even the punch line: my little brother is either dead or not. This point is not clear.

SIX

Byron isn't dead. He's got to be hurting, though, propped up like a stuffed owl against a hospital pillow. Some numskull of a nurse has brushed his hair up into a peak. I don't draw his attention to it by flattening it down. But my hand aches to go for my comb. Byron's right arm is in a sling, making an acute angle. Bandages run over his shoulder, partway up his neck, and in the other direction around his rib cage. He's trussed up and probably itching. I don't know what kind of medication is keeping him this calm. Maybe none.

This is the morning of the second day. I never exactly sorted out the first day. The message, it turns out, came from Grandmother *at* the hospital. And Byron was in the emergency room, not the morgue. Nobody actually mentioned morgue, but on the West Side IRT downtown express I have sharp impressions of cold slabs and clanging metal drawers. I am in a bad way.

When I get to the hospital that first day, Mrs. Schermerhorn's Cadillac is parked in the "Ambulance Only" zone. But Nathan's already taken Grandmother home and has come back, bringing Mr. Kirby and waiting for Mrs. Schermerhorn. They've already set up their routine. Dammit, you've got to hand it to those Gray Panthers.

Byron has maybe five blocks to walk home from school. He walks halfway with his buddy, Tim Somers, who cuts off down Garden Place. So we're talking about two and a half blocks alone. He doesn't make it on this particular afternoon. At the Joralemon–Henry Street intersection he's mugged by a gang. They pile out of an old Chevy without plates, driven by an older gang member who may or may not be the honcho. They wipe the crosswalk with Byron. Stomp him. Slap him around. Take his billfold. Shake him down. Then when they've picked him clean of a dollar and change, they kick him flat in the middle of the street, briefly blocking traffic. And some hero gives his arm an extra twist which in some freak way breaks his collarbone. "Right clavicle," to quote Blue Cross/Blue Shield.

Plenty of witnesses, of course. This is New York. A woman carrying a baby and a bag from Key Food somehow gets him out of the street after the Chevy tools off. The doorman of some building calls an ambulance, which takes its time. But nobody sees anything. This is New York.

The head night nurse, who pretends to speak no English, puts me bodily out at the end of visiting hours. But I'm back the next morning. The academic life of Van Cortlandt Academy can struggle along without me. Likewise the Giants and Dwarfs. Let them think I'm having a delayed reaction to Mom's death. In a weird way I am. This is hitting me harder. I'm less numb, but that's only part of it. I'm feeling this more, and I'm not even guilty about my priorities. Even when I see Byron sitting up in the morning, coolly examining the sling like it isn't even on his arm. He has unbelievable detachment.

I try to read his eyes—a pointless habit of mine. *Where were you when I needed you?* No, I don't see that. It doesn't figure anyway. You shouldn't have to walk an eight-year-old kid home from school. On the other hand, that may be the only way to get him home alive.

"How's it going?" I say for a brilliant opener.

"It's broken," he says, sounding impressed.

"I meant—how's it going generally? You feel okay?"

"They have oatmeal for breakfast here too," he says. This is when I itch to smooth down his hair. In another way, I'm afraid to touch him. His unwrapped shoulder is just this small white mound, half hanging out of the

[65]

ridiculous nightshirt thing they've got him wearing.

"You sleep any?"

He nods. I'm glad he doesn't ask me the same.

"Want to talk?"

"What about?"

Hell, I don't know. Who do you like in the All-Star Game? Read any good books lately? What do you want to be if they let you grow up? "About what happened."

"I don't want to talk about that," he says, shaking his head. "Definitely not."

"Don't you want them to catch those . . . turkeys? Don't you want—" "Justice" is what I'm trying to say. But how do you ask an eight-year-old kid if he wants justice? Particularly when there isn't much of it around.

"I don't want to talk about it to *anybody*," Byron says. His head never stops shaking. "Come closer, and I'll tell you why." My knees are already nudging his bed. He wants me to lean over so he can whisper. He doesn't know if there's somebody in the bed on the other side of the curtain or not. There isn't. Even so, he wants to whisper.

"There were girls with them. In the gang." He lays his one small workable hand on the sheet and stares at it, while I think this over.

"Girls?"

"Yeah. It was a girl who took my billfold. I think it was a girl who stood on my back too. When they grabbed my arm and—"

"Okay. You don't have to go through it." Who am I

[66]

crying to spare here—him or me? He has heel-mark lacerations on his back which I haven't seen yet. I'm still not getting his meaning, and he's a little exasperated about how thick my head is.

"I don't want anybody knowing girls were in on it. How'd you like it if people said *you* got mugged by girls?"

Macho in the third grade. I can't believe it. "But there were girls *and* guys, weren't there?"

He nods. "But when people start talking, pretty soon it'll be girls only. Anyway, I couldn't identify them. And what if I could? They wouldn't do anything to them. They'd probably all be in Juvenile Court, which is nothing. Except maybe the driver, and he never touched me."

Oh, Lord, he's got it all worked out. Even the revolving-door legal setup of the Brooklyn Family Court. Why do you have to grow up so fast? Why do you have to know all this just because it happens to be true? And what am I supposed to do about it all?

"So I'm not talking to anybody," he says. And I see the line of Grandmother's jaw, miniaturized, when he stops talking. It's that little touch of inherited toughness, which I didn't inherit, that puts me over the edge. I pull back from him and start away. The tears that have been hanging around under my eyelids for weeks are back in full force, ready to let go. Always ready at the wrong time. "You going to school?" Byron asks.

"No," I say to the curtain, "I'm not going to the

[67]

dumb-ass Van Cortlandt Academy today." I can feel his awed look through the back of my head. He thinks my school is Yale University, but classier. "I'm going down to the cafeteria for coffee. I'll be back in a while. You want me to bring you anything?"

"I guess not," he says in a thoughtful voice, psyching me out.

But I never make it to the cafeteria, where I wasn't heading anyway. I was planning on going down to the waiting room, which is full of Spanish-speaking people who know the value of tears and who cry if they want to, in unison and in groups. And there I'm going to let go among strangers. And I am going to cry my damn eyes out. Except I don't get there either.

A couple of steps from Byron's door I'm face to face with this tall, skeletal guy with a paunch and thin hair. I know him from somewhere, way back. And just as he asks, "Is this Brian Atwater's room?" I place him. This is the headmaster of Brooklyn Heights Collegiate. I remember him because I went to the lower school there a year when we first came to live at Grandmother's. And he hasn't changed. I have, of course, so I say, "I'm Jim Atwater, his brother."

"How is Brian?"

I let the wrong name pass once, not twice. "His name's Byron. Broken right clavicle. His arm's in a sling. He won't be writing any essays for the rest of this term."

This tells him I know he's from Byron's school. Still,

he doesn't know me, so he says, "I'm Brewster Stewart, the headmaster."

We shake hands. "Can I see . . . Byron?"

"Sure. He'd like that." Then something makes me add, "Do you know Byron?" even though I have the answer to this one already nailed down.

"Well, I don't exactly know him," Headmaster Stewart says. "I can't know them all, can I?" I let the answer to this one dangle. I ought to be glad he's bothered to come around, but somehow this doesn't cut much ice. "But maybe I could have a word with you first," he says.

I steer him away from the waiting room, and we find a couple of plastic chairs at the end of the hall. He plants enormous hands on his knees and clears his throat. You don't make Headmaster without public-speaking skills. "I understand that the boy—you boys—have recently lost your mother." He sidesteps the topic of our father. It's all there in the school records, which he's speed-read in preparation for this visit. "We were notified of Byron's accident by your grandmother, Mrs. Schermerhorn."

"Our grandmother is Mrs. Livingston. Mrs. Schermerhorn's a friend of Grandmother's. Makes calls for her, things like that."

Stewart absorbs this, probably getting a mental picture of Grandmother as senile and helpless. Ha.

"And it was no accident. He was attacked by a gang whose methods sound reasonably professional. They beat him, robbed him, and then broke his collarbone. When you see him, you can imagine what kind of a fight he could put up. I think there were seven or eight of them, and they had wheels."

Stewart pulls thoughtfully on one long earlobe. "I see," he says. "Police called in?"

I think of Byron, then Grandmother, then the entire Kirby/Monday Evening Club power elite, complete with low-profile legal counsel. "Probably not."

Stewart gives this some thought. "And where did the . . . incident take place?"

"In the middle of Joralemon and Henry. He'd have had a better shot at crossing against the light in front of a truck."

"Ah, well," Stewart says, leaving his ear alone, "it was well away from school property then."

Even before I can work on the meaning of this—and the question about the police—I'm beginning to tense up. Stewart's on his feet, ready to walk away. "I'll just stop in and say hello to Byron anyway. He has an excellent academic record, as I expect you know."

"I hope being mugged won't count against him, especially since it wasn't on school property."

This turns him around. "Was that an attempt at sarcasm?"

"To tell you the truth," I say, "I'm not sure. I can't quite figure why you're here except to make certain you're in the clear."

"I have a responsibility to all our students while they are within the school's jurisdiction. If you were older, you might be able to fathom the legal ramifications. As it is . . ." He gestures in the air, waving me out of existence. I decide not to go.

"Even without the jurisdiction and ramifications bit, though, it wouldn't look too good if word got around that Brooklyn Heights Collegiate kids are being creamed on the way home from school. Might cost you a little credibility, and tuition money."

"I can fully understand that you must be upset about your brother's incident."

"Will you do me a favor and stop calling it an incident? 'Incident' plays it down too much, even for you." I'm beginning to roll now, even with no history of getting mouthy with headmasters. "If you feel any responsibility to your kids—if you can feel anything—why don't *you* call in the police?"

Stewart takes a deep, sighing breath, calls on Heaven for patience in dealing with this maladjusted, deeply troubled lad giving him a hard time. "Surely calling in the police in an inci—crime committed on a public street involving a minor is the parents' responsibility."

"But we're running a little short on parents," I fire back. "And as for my grandmother, Mrs. *Livingston* . . ." I let that go, let him remember his impression of Grandmother as a drooling old vegetable.

"As you point out, I'm just a kid myself. It looks like

[71]

you're the logical one to call in the cops. Who knows, this could be the start of a tremendous crime wave. They could be picking off your students like flies. And as for the hospital taking any action, you can forget that. They do well to get the bedpans circulated—at a hundred bucks a day."

Where I'm getting this line of argument from I don't know. In a way I don't even want this yo-yo in a vest calling in the police because Byron doesn't want it. But I like seeing Stewart stew. It's making my day, and I haven't had what you'd call a good day in quite some time.

"Young man, what school do you go to?"

"What's that got to do with anything?"

"Just about everything. As you're not a student at Brooklyn Heights Collegiate, I'm under no obligation to you, legal or otherwise."

"For all you know, I *do* go to Collegiate. You wouldn't know your own students unless they fell behind in their goddam fees." I'm letting my voice get out of control, and it's echoing down the corridor, cutting down on the effect. But I'm through now, and Stewart's not going to walk away from me. I push past him. But before I get as far as Byron's door, I turn around. Stewart's still waiting for me to vanish. "And another thing," I say, loud and clear, "you can forget about that little visit to my brother's bedside. He can spot a phony every time."

I blunder right into the middle of a bunch of interns

with clipboards, making the rounds with a real doctor. They step aside to make way for this crazy kid I am, manifesting early signs of hysteria, paranoia, whatever. And when they clear a little space around me, I see the tall, stooped figure of Mr. Carlisle Kirby stock-still behind them. I'm louder than his deafness. This is perfectly clear. He trains on me an old, watery, despairing eye. I lurch around him and head off down the hall, not knowing where I'm going. And no longer near tears. Tears? What do I need tears for?

"You feeding Nub?" Byron asks me that night. I forget where the day went. It's evening visiting hours. One dim shaded lamp next to Byron's pillow, and shadows. I've pulled the curtain back so he can see we have the hospital room to ourselves. We're alone all right. The Gray Panthers have decided to make themselves scarce as long as I'm keeping myself on an around-the-clock call. Mr. Kirby isn't going to be caught out in public with an unstable, yell-prone kid. Mrs. Schermerhorn has sent the usual bunch of red, white, and blue carnations. She's enclosed her calling card, engraved by Tiffany. No cutesy message of get-well-quick written on it. Cutesy messages are not her style, if she has one.

Am I feeding Nub? That is a tough question. I forget about Nub for long stretches because Nub's two greatest talents are keeping to himself and sleeping.

Byron says, "Nub eats a half can of Gourmet Feast in

the morning and a soft pack of Tender Vittles at night.
He snacks on Kat Krunchies, so I keep a bowl full for
him."

"Ah . . . ," I say, "I'll look into it. Almah's probably feeding him." On the other hand, Almah's probably
not feeding any ex-stray who sheds all over the carpets.

"If she's feeding him on table scraps"—Byron sighs—
"he'll start throwing up in corners."

Almah has no table scraps, ever, as we both know.

"Well," Byron says with the calm that passes all understanding, "there's plenty Gourmet Feast and Tender
Vittles and Kat Krunchies. I always keep a good supply."

No direct request that I rush home and feed Nub. No
whining, no wringing a promise out of me that I won't
let old Nub starve. Nothing so easy. There's so much of
Grandmother in this little kid that I feel outnumbered.

"I'll feed Nub." An even greater calm falls all over the
hospital room. "I'll let him into your room so he can
sleep on your bed if he wants to." I want to promise
other, more fantastic things on Nub's behalf. Freshly
caught mice, hand filleted, a rhinestone flea collar, an
extra distemper shot for complete health coverage. I let
it go with the original offer. And all this time Byron is
thinking. I feel little sonar beams of intense thought
radiating in this peeling green room. Byron's moved on
from Nub. Like I say, he doesn't harp or whine. Also, he
doesn't ask when he can go home from the hospital.
After all, why should he be in any hurry to return to the
outside world?

[74]

"Remember how we used to sit on the stairs Monday nights sometimes?"

This comes out of left field, of course. I'm not prepared, but I know what he's talking about. He means him and me and Mom, and how we used to creep half-way down the front stairs on Monday Evening Club night and listen to the oldsters conducting their session.

Mom wrapped up in her bathrobe with cream all over her face, and Byron and I edging very quietly down as far as the top of the parlor doorway, like three kids eavesdropping on the adults. We had a whole routine worked out about avoiding the stairs that creaked and finding the right level where we sat in a row to hear Mr. Carlisle Kirby chairing the meeting: asking Mrs. Schermerhorn to read the minutes, calling on Mrs. Garret Pierrepont to give the financial report, making a small joke about how Mrs. Dykstra is to maintain order because she's sergeant-at-arms. Most of the meeting's taken up with Sanka and Robert's Rules of Order.

I remember how Mom would fold her bathrobe under herself as we silently settled, and how our elbows touched as we sat there. Nudging each other, trying to crack each other up without letting ourselves succeed. I can remember Mom's presence between us, her hands clasped around her knees. She was tall, like Lorraine, but more graceful. I remember her hands, the length of her fingers, the tapering nails. But I don't try for any more personal details. After all, we were sitting there together in the dark.

[75]

Why is Byron remembering this? It's been months since Mom felt well enough to sit on the hard step in the drafty hall. "Sure I remember," I tell him.

"You think Grandmother knew we spied on her meetings?"

I'm sure she did. She probably even wanted us to carry on the good work of the Monday Evening Club in our own time.

"I can remember way back," Byron says, this old duffer recalling his youth. "But I couldn't figure out why they were always trying to do something for that one man, that Mr. Nixon."

This calls for an answer far more complicated than cat food. It's weird too because the subject of Mr. Nixon is practically the only topic Grandmother's ever discussed with me. And she only raised it once. "I don't know what they're telling you in school, if they're telling you anything," she said to me, "but it remains very fashionable to excoriate President Nixon. I trust you are aware of this, Jim."

I was—barely—and this was also the moment when I learned the word "excoriate," an excellent Grandmother-type word.

"He remains a favorite whipping boy for a nation that has lost a crucial war. The Vietnam War," she adds, leaving nothing to chance. Of course she's right. We learn nothing about it in school except that Nixon was a shifty liar who broke into hotel rooms and would bug your phone before he was caught and sent away to San

Clemente, which may or may not be a foreign country.

Grandmother sees that this is not a hot issue with me. "I have no intention of compensating for all the distortions and omissions of your schooling, Jim. If I did, there wouldn't be time for anything else. Mr. Nixon ended the war, and whether that was a good thing or not remains to be seen. But it was a very bad thing for his enemies: the Pentagon, the arms manufacturers, the labor unions, the shipping magnates. A great many powerful people were profiting by the Vietnam War. When President Nixon brought it to an end, his enemies took their revenge. He had far worse enemies in his own camp than in Peking or Hanoi."

I get a little delirious here because I'm rotten at geography. I'm not red-hot at history either, if that's what this is we're talking about.

"I have never tried to form your opinion upon any subject," Grandmother says in this long-ago session we have. "But there is something I want you to remember. I won't mention it again. President Nixon ended the draft—after more than thirty years of forced military conscription in this country. This means that unlike earlier generations of young men, you won't have to face two or more years of obligatory army life, or leave the country to avoid it. This gives you back two years that belong to you, and perhaps much more. You can pursue an education, a career, without interruption. President Nixon gave you your freedom."

I get a sudden flash of ex-President Nixon as a bronze

[77]

statue in Borough Hall Park, freeing the slaves. I've never heard Grandmother on a soapbox before or after. Even at the Monday Evening Club session you never hear her voice except when she leads the Pledge of Allegiance. She's this mute, impressive presence in the wing chair who could take the floor any time she wants to, and so doesn't.

"Of course," she adds, "the draft will be re-instituted in the future. It's too lucrative for a great many faceless, nameless people who will gladly grow fat at the expense of young people's time and lives. And they will bring back the draft in some future Democratic administration. I can only hope you will be beyond the age of vulnerability. I hope they won't make you do their dirty work, Jim."

This is as tender as I ever remember Grandmother being, the way her look stayed on me for a moment. Then she turned on her cane, and that was all there was to it.

Byron is still waiting for a clear, sensible answer about Richard M. Nixon. "I think they figure that politics is so full of crooks that one particular one shouldn't be singled out." This is about as puny an answer as I have ever given anybody on anything.

"I don't know," Byron says, rubbing under his nose. "I don't think they'd send him birthday telegrams if they thought he was a crook."

How true, how true. But this isn't to be the major

topic of the evening. Byron lets the whole subject drop. Scratch the Fiend of San Clemente and his faithful friends in the Monday Evening Club. And Byron's had the last word on the subject. Shades of Grandmother again.

What he's doing is leading me carefully back into the past. "What do you remember about way back?" he says. "About Dad?" This is the whammy we've been heading for through the Nixon smoke screen.

Okay, what do I remember about Dad? Not much. I was nine when he went, and less mature than Byron is at eight. I remember . . . what? Things you can picture that don't translate out in words. Swinging on Dad's belt buckle for one thing. This goes back to the dawn of time. But I remember riding the flume at Six Flags outside St. Louis with him, wedged into this fake log with seats, shooting artificial rapids, hanging on for dear life, hanging onto him. I can remember cooking out at a campsite, sleeping out, with Lorraine and Mom and Dad and me in sleeping bags. I remember Christmases—stuff like that.

But I can't replay all this for Byron. Because they're all happy memories. And he doesn't remember any of them. And why should he have to wear my hand-me-down memories? They wouldn't fit.

So I lead him astray. "This is after Dad, but did I ever tell you about the time Lorraine saved me from the killer dog?" Byron looks at me. He never had much of a relationship with Lorraine either. She was too much older. Also, he's heard the dog story before.

"Tell it," he says, willing to humor me.

I jump into the story and try to remember how it goes because he'll remember details from before. "Well, I was about ten—eleven tops. Lorraine was in college. This is before she dropped out and got married." My sister Lorraine doesn't look like any of us. She's a big, towering, thick-legged girl. I always remember her in jeans and sweat shirts, a walking offense to Grandmother. In other ways, she and I are somewhat alike, emotions kind of close to the surface, with uncertain control.

"One night it was really hot," I tell Byron. "And Lorraine and I walk over to Baskin-Robbins for ice cream. Anyway, we're coming back down Hicks Street when toward us comes this little old woman with a big dog pulling at the end of a chain-link lead. I'm not sure of the breed now—make it a Doberman. Definitely an attack dog. He gets up close to us and notices those ice-cream cones. He makes a lunge at me. I don't have time to get worried. Suddenly this big paw is splayed out across my chest, and all I see is this enormous slobbering mouth. The top dip on my cone hits the sidewalk. The big Doberman drops down right away and starts lapping it up. He has a tremendous taste for fudge ripple.

"This is no big thing. Even if the dog's rabid, we're practically home free while he licks all the way around the ice cream, not even coming back for the second dip.

But there's this little old woman at the other end of the leash. She looks really worried for a minute. Then she yells out, 'Leave my dog alone! He never bothers strangers unless they bother him first!'

"I thought this was pretty astounding. I wouldn't have later on because I saw her around a lot, being dragged by that dog along every street in the Heights. And yelling at people who aren't even there.

"But this drives Lorraine crazy. She starts quivering like jelly and says to this old woman, 'Listen, you old fool, you better get that dog off the street. He attacked my brother, and I'm going to have the police in on this, and you'll be back in a padded cell before you know where you are!' I've never heard so much hate in a human voice. Never.

"The old woman blinks like Lorraine is the insane one. She's used to people avoiding her, of course. 'Police?' she screams out. 'We'll see who has the police! That snot-nosed kid went for my dog! I seen him! Police! Help! Somebody get the police over here!'

"Of course nobody did. And the Doberman and I are just standing around waiting. And I'm still reasonably sure that Lorraine and the crazy woman are going to start mixing it up with fists and hatpins and whatever. The Doberman keeps lapping up my ice cream, making long tongue strokes on the pavement. And the woman goes on raving.

"Then Lorraine reaches down and seems to be going

[81]

to rub the dog under the neck. But she unfastens the leash from his collar instead. Of course he just goes on eating my ice cream.

"And Lorraine swings her foot around sideways and connects with his rump, hard. The dog gives a startled jump and realizes he's free. So he lights out between two parked cars and sprints up Hicks Street. He's almost out of sight before the old woman sees she's holding a limp leash.

" 'Oh, my Gawd,' she whispers. 'Oh, my Gawd in heaven!' She darts out into the street—screaming, carrying on. It's really awful—begging her dog to come back, though he's already around the corner of the Bossert Hotel and probably traveling hard.

"He must have come back because I saw them around later. But the way Lorraine acted shakes me up at the time. 'What if he's gone for good?' I say to her. 'The dog, I mean.'

" 'Better for him if he is,' she says. 'If I see him again, I'll kill him.' " This is old nonviolent, sentimental Lorraine talking.

Byron listens to this retold tale with his usual patience. In fact, more. He's staring off into space past my shoulder. If he didn't look so alert, I'd think I'd talked him into a trance. Besides, it's not that good a story. For one thing, it's about Lorraine and me, so it leaves Byron out. For another, there's a parallel here that strikes me in the retelling. A parallel which I'm sure Byron grasps. I felt smothered by all Lorraine's fierce protectiveness. She was slopping over with love—there's no other way to put it. And because of this she was leaning on me hard. Maybe she felt Dad's leaving worse because she was older and because she was a girl. Maybe it was just this emotional thing with her. I was glad when she got married and left, though I never admitted it to myself.

And what if I'm doing the same thing to Byron—smothering him? I decide fast that there's no parallel. I have to look out for him because he doesn't actually have anybody else. And I don't fight all his battles for him, do I? Otherwise, would he be in this hospital bed?

Time passes while I'm dealing with myself over this. But Byron's staring over my shoulder. If he didn't blink once in a while, I'd be worried.

Visiting hours are about over. The warning bell's rung. Footsteps in the corridor. Doors creaking. Byron blinks again. At last I turn around, follow his gaze. And I see there's somebody inside the door, standing there, maybe for some time. Enjoying this wonderful saga about Lorraine over-reacting to a dog. But this is somebody who hasn't heard the story before.

This is—it can't be, but it is—this is Dad standing there. In the same rumpled suit from the funeral, with a flight bag in his hand. Dad looking travel-worn, as if he'd been suddenly sent for, which he has.

SEVEN

The flight attendant—a gorgeous blonde in a snappy wrapper printed all over with the airline logo—gives Byron's slinged arm a look, then another at Nub in his cat carrier. Then back to Byron. I've buttoned a short-sleeved shirt of my own around him, with room inside for his angled elbow beneath the shirt sleeve that flaps pathetically. For days now he hasn't let me feed him. He's learned how to eat and perform other personal functions with his left hand. Of course the kid turns out to have latent ambidextrous tendencies.

"Listen, chief," she says, "I don't know how you got that animal all the way to the aircraft, but it's supposed to be checked with the baggage. No pets in the passenger compartments."

This is absolutely no complication. Byron already has one foot on the plane. I'm behind him, and behind me is a long line of Florida-bound pilgrims snaking all the way back to the boarding lounge. Maimed Byron just looks up at her with solemn, sensitive eyes, not even acting. "Oh, well," says Miss America of the Airways, "you keep—whatever it is—in its cage, okay?" She reaches out to pat Byron, but he's all elbow and eyes. So she grins him past her. Her head bobs a little—let's live dangerously on this flight. Rules are made to be broken.

Score one for Byron. Nub travels with us. Pets have been known to depressurize suddenly in the baggage hold anyway, and the blonde knows this. Besides, it's the arm in the sling that blows her away. There could be a large gray rat with rabies in the carrying case. Nub can't be bothered to push his whiskered face against the screen-wire window. He's sitting in his miniature R.V. with his tail slapped against the opening.

We've gotten Nub past check-in and through the passenger shakedown for concealed weaponry, courtesy of Mr. Carlisle Kirby, who's seen us off. While he blustered at the checkpoint, Nub rode right through the X-ray machine like a piece of hand luggage, which he is. Being X-rayed to make sure he isn't a cat-shaped bomb will probably give him some fatal disease, but he's still with us.

We're already working down the aisle of the Florida plane, looking for our seat assignments. I let Byron carry Nub's case in his good hand. There are more attendants on this plane, which has three seats on each side of the aisle. All the window seats are taken up by big-gutted businessmen who won't even look at the view. Oh, Lord, if I am ever a middle-aged businessman, don't give me a fat gut. Make me jog. Make me steam. Make me work out till I drop.

So Byron won't even see out the window. We settle into aisle seats opposite each other, and Byron shoves Nub's case under the seat ahead of him. We're strapped down and ready for blast-off.

I have only myself to blame for this whole thing. In the countdown to summer, we were measuring in days. We could have moved from schools to camps without any trouble to anybody, especially ourselves. Even Byron's arm wasn't a major drawback.

No, I'm the well-known straw that breaks the camel's back because of my explosion at Byron's headmaster— badly timed and witnessed by Mr. Kirby. The Senior Citizen jungle telegraph lights up like a switchboard. Drums throb in the Brooklyn Heights wilderness. Word gets back to Grandmother, who's known all along that one or both of us will get out of hand one of these days. And I have. I've trodden lightly around Grandmother for many years, and I have blown it in one minute. Now that I've started rebelling against authority, it could develop into a habit. It's true. It could. It felt good.

And so my dad is summoned from Florida. He flies up and stays overnight at the Bossert Hotel. Grandmother doesn't run a rooming house for ex-sons-in-law.

Byron's injury is the excuse, but Grandmother has a private audience with Dad, and I know I'm the big topic. Dad goes back to Florida on a morning flight, still shying from any direct confrontation with Byron and me.

Later Grandmother calls me into the middle parlor to give me fair warning of what's coming off. She's in the wing chair, with her Lucite cane tucked beneath. The strain of having encounters with both Dad and me in the same week is telling on her. She looks old, tired, still grieving for Mom, and I feel responsible for every wrinkle. Believe it or not, I love this old bird. It's easy to love her—she doesn't expect it.

This is not a good scene for her. In all her dealings, she has to be in charge, so the burden's on her. I never open my mouth. I have already opened my mouth—thus, this scene.

"I reared your mother," she says, setting things up. "You may have concluded that I—resent her for . . . the way she died."

I haven't thought about it particularly, haven't had to. It's too obvious. In Grandmother's private universe you suffer anything. And you never cheat. Not even death.

"I would have resented her dying anyway. I was prepared to." Her lips come together in a line above that rock jaw. It's easier to talk even about this than to come to terms with my case.

"I reared my daughter—your mother—and gave her

[87]

my best. When she came back home with you three children, I took her in. I don't believe in divorce, but I wouldn't have turned you away from my door. I am not as stiff-necked as you suppose." Here she gives me a look that's almost an invitation of some kind. But I don't risk a response. She already has me between two bases with the ball in her hand.

Besides, I'm thinking too hard, though not exactly along her lines: someday this old lady will die too, the last of her breed, frozen in her pose. And one of the few encounters I'll have left to remember is this moment. I'll think back on this even if it hurts.

"Your sister, Lorraine, was a trial to me," she's saying.

How true, how true. Klutzy Lorraine who cried over old TV movies. Aimless Lorraine who studied Ed. Psych. and married a sergeant she picked up at a lunch counter. I have a crazed mental flash of big Lorraine sweeping up a small sergeant in her arms and trying to carry him out through the revolving door at Chock full o'Nuts.

"I could never see allowing a girl simply to follow her own inclinations. And Lorraine had so few." Grandmother isn't even on the subject of Lorraine for Lorraine's sake. She's using her to get across the idea that she's already raised her own child; she shouldn't be expected to raise another whole flock—especially us. The gap's too wide now. I see this. I almost want to nod.

"Of course you boys were an even greater problem to me, especially after your mother was ill and I couldn't be sure how much strength and judgment she had left for dealing with you."

Strength and judgment. Prime Grandmother-type words. A whole creed.

"I have tried to know what was best since . . . in these past weeks." Grandmother stops. Her fingers fidget on the chair arms. Something crosses her face. "If only you were both *older*," she says, "just a few years *older*." I can really feel for her. That's all I want myself. It's eerie how close we are and how far apart.

"If I could give you years of my own, I would!" she says in a voice suddenly so strong that it makes the skin on the back of my neck prickle. Then her voice is quiet again. "I'm going to send you down to your father in Florida."

My pupils must be dilating. I didn't see this coming.

"I can't manage, and I don't take on responsibilities I cannot discharge. You can come back, of course, but not until the end of summer. Then, perhaps, I . . ." She starts to say something about herself, about how maybe she'll be further from Mom's death and closer to her own idea of "managing." She switches away from herself in time. "You and Byron have your schools, your friends here. . . ." She looks around in her mind for other things we have, finds nothing, and lets it go at that. "You both need a father. Whether or not you can develop a relationship with him is out of my hands. I don't know what your memories of him are. They will mean very little now. You will not be starting over. You will be starting fresh."

Not a word against Dad, nothing about his strength and judgment, if any. Grandmother's never been seen carrying a grudge. Dad was never her responsibility.

And so we're off to Florida, with no back talk. And Dad is about to become a father.

I don't let myself think about it. I don't even tell Kit Klein I'm not going to take the counseling job in Vermont. I can't believe this is happening. I'm vanishing without a trace into off-season Florida with one somewhat crippled-up brother, an unresponsive gray cat, return tickets, enough savings money to make us all independent, and a canvas bag full of shorts and Lacoste shirts.

Byron's good arm is clamped on the chair arm, and his head's out in the aisle. He seems to be counting the house. The flight attendants are whisking late arrivals into their seats and taking drink orders. Byron's bright as a button. This is a lot better than Camp Arrow Rock. I'd like to think he'd rather spend the summer with me instead of without, but this point hasn't come up. He doesn't seem worried about an entire summer with an unknown father with a lousy track record for reliability. And for this I'm glad because I'm worried as hell.

After a long delay, our number comes up on the runway, and we take off. This is a tremendous kick for Byron even though he can't see anything. He's playing pilot. He has the keen look of Snoopy trying to fly his doghouse.

Somewhere over New Jersey this girl comes up the aisle from behind us. I've already spotted her in the boarding lounge. She's easy to spot. She steps over Byron and settles into the middle seat beside him. His head pivots, and he gives me a penetrating look. His entire face is trying to tell me something. But I'm not too attentive because I can see the girl in profile over his head. She's almost a knockout, with auburn hair still electric from the brush. She's wearing some kind of sundress and has this great scattering of freckles all over her shoulder. She reaches down to put a small case under the seat, next to Nub, and the light catches that auburn hair. Naturally she's not sitting next to me because this is the kind of luck I'm cursed with.

Byron's still staring at me, and his mouth's making strange shapes. His eyebrows are meeting over his nose. What's the matter with this kid? If he wants to swap seats, he's got himself a deal. I look at the girl again. She looks like an ad for something forbidden. She's also about my age, though I am willing to be any age she likes.

Byron's eyes are darting in every direction. He can't lean very far out into the aisle because of his seat belt, but he heaves over my way, points at something, and whispers, "Give it to me. Quick."

He's pointing to an old boarding pass somebody left behind in the seat pocket ahead of me. I see his own pass sticking out of his sling. Maybe he's making a collection. "You want this?"

"Yes," he hisses, "but keep quiet." What's he up to? His whole mood is total cloak-and-dagger. I hand the pass over. More mystery. He pulls his own boarding pass out of his sling, replaces it with the old one, and then gives the girl an almighty nudge. She's turned full-face now: a true knockout. Her eyes widen, and she listens some more. She seems to be taking the boarding pass, plants it prominently in the seat pocket in front of her. But Byron's talking on and on. She gets in a few words, but very few. What a line this kid must have.

A stewardess crosses my line of vision. When next seen, Byron's staring straight ahead, looking innocent and a little bored. But pretty soon he's whispering to the girl again. And she's hunching over, whispering back. I watch the perfectly straight part of her auburn hair, and I still don't understand any of this. When she sits up again, she gives Byron a look I've seen people give him before, people who think he may be a midget disguised as a kid.

He turns to me and says, "I have to go to the rest room."

"So go."

"I can't manage by myself," he says, waving his slinged arm. This much I can grasp. He wants me to retire to the rear of the plane for a powwow. We troop down the aisle, and he stops dead in front of two vacant rest rooms.

"Bend over," he says, "I can't talk out loud." Then I get the story. "She doesn't have a ticket," he whispers, wetting my ear. "She's a *stowaway*."

"Can't be."

"Is."

I know it can't be. I saw her getting her seat assignment in a legal manner. Still, Byron's getting a tremendous kick out of this. I play along. "But how do you know?"

"She finally confessed. When I gave her my boarding pass. This may not solve everything," he says, shaking his head. "They may make a head count."

"Wait a minute," I say. "You thought this up—out before she . . . confessed."

"Sure. I watched when everybody came on. I'm pretty sure she wasn't with the rest of us. So she had to be locked back here in the *toilet*." He points at a rest-room door in case I've forgotten what a toilet is.

I concentrate on keeping a straight face. "How come you gave her your boarding pass? Why didn't you give her the old one?"

He tries to be patient. "In case they look close, I have a *ticket*. She *doesn't*. Besides, they saw me and Nub come on board."

I can be as sober as he is. "You know this is illegal, of course."

"She has to get to Florida," he explains.

"Is she a runaway?"

"I don't know," he says. "Maybe." The idea appeals to him. "Probably."

"How does she pull this scam?"

"She didn't want to tell me, but finally she did. She knows the airport like the palm of her hand." He holds out his palm. "She knows how to board just before the flight crew does. She's done it before. She didn't want to tell, but finally she did. One time they caught her, but they didn't do anything to her. She pretended she was epileptic."

We're both thinking this is a very inventive girl, but not for the same reasons. I stare at him, not wanting to spoil this wonderful trip he's on all by himself. "You know," he says, "epileptic." His tongue lolls out, and his eyes roll around.

"Oh, good grief," I say. "Let's swap seats. I want to talk to her."

"I'm telling you exactly what she admitted," he says.

"There are other reasons for wanting to sit next to her."

We troop back up the aisle. Byron says over his shoulder, "Her name's Adele Parker." I hope after he's gone through puberty he still has his talents. He's had her eating out of his hand.

"You sure that might not be her code name?"

He shrugs, frowns, adds this to his profile of her. I ease into his old seat. Byron flops into mine and becomes instantly wrapped up in "Emergency Instructions for the Boeing 727." He's going to cover for us. "Well," I say, "how's it going, Adele?" I figure Byron's broken the ice in his own unique way.

"A little strange so far," she says, slanting a look my way. "Is he really your brother?"

"Really."

"He's dynamite."

"He is that. His name's Byron."

"Byron?" Adele tilts her head. Her nose comes to a perfect point, history's most beautiful bird dog. "Byron's such an old-man's name."

"Therefore it fits him," I say.

"He's got an unbelievable imagination," she says. She's not totally convinced of my own grasp on reality. "They had me back in the smoking section, which I can't stand, so I moved up. And just to keep the record straight. . ." She has an open-topped carryall in her lap. She parts the top with two fingers, and there's her plane ticket with the rim of her own boarding pass stuck in the folder. We both look at it in silence while Byron across the aisle is reading every word about the Boeing 727. He's whipped the system on behalf of a damsel in distress with larceny in her heart, and he is very pleased with himself.

"Listen," I murmur, "thanks for not putting him down. Did you actually tell him you have an epileptic act?"

Her tongue lolls out, and her eyes roll around. "Okay, okay." I glance over at Byron. He's caught this and nods with a quick I-told-you-so look. Strung up between the two of them, I'm somewhere between cracking up and wanting to make an all-out play for Adele Parker. Naturally, I don't do either. Something stirs, bumps, and hisses under my feet.

"What *is* that?" Adele draws up her feet; her eyes go wide again. They're green with gray flecks in them. Her eyebrows are almost invisible.

"Oh, that's our cat," I say, somewhat lamely.

"Speaking of stowaways." Adele looks heavenward. The attendants are racing up and down the aisle, bringing lunch. "Does he watch a lot of TV?" Adele asks.

"No," I say, "he just sleeps, sheds, and uses his litter box a lot."

"I mean *Byron*," she says.

My face is suddenly on fire. Why is it every time I'm around a girl, my I.Q. drops a quick fifty points?

"I mean he worked up an instant plot about me smuggling myself onto a plane in about two seconds. It sounds like TV."

"No." I struggle to say something that makes sense for a change. "He never watches. You see him in his natural state. He doesn't need TV."

Lunch comes. I plan to keep the conversational patter going. This is the time when discussions can die, while you're struggling to get the plastic fork ripped out of its cellophane sack. Besides, I have this need to change the general image Adele's getting. Spacy little brother,

[96]

bumbling big brother, cat traveling economy—in short, a zoo.

But I notice that Byron's starting with his banana pudding, bypassing the chicken and green beans. "Leave the dessert till last," I tell him across the aisle.

"You really do the big-brother thing, don't you?" Adele says, starting on her salad, in case I start ordering her around.

"Not especially," I say, too quick. She's probably been bred not to call people liars. But I'm talking away like crazy now. Words flow out of me, even a late introduction. I tell her we're going to spend the summer with Dad. She's going to Miami to spend the summer with her mother. Her father in New York has custody of her, but her mother has these vacation-visit rights. I skip over the fact that Mom's dead. Adele gets the idea that our parents are divorced too. This is near enough the truth; I'm not looking for sympathy. What I'm looking for I don't try to picture. The airways seem filled with the children of broken homes, migrating like birds.

She goes to Spence—a very statusy-type school—my year, senior-to-be. I've never been elbow-to elbow with a Spence girl before, even at a mixer. She knows Kit Klein through a friend of hers who lives in his building. I'm not sure whether she knows him from his pre-couth, tough-shitsky days or not, so we don't linger over this.

It comes out that I've never been to Florida before. "What's it like?" I ask her.

She thinks. A tiny silver chain around her neck
nestles in the hollow of her throat. Beyond her, past a
fat businessman, the clouds stand still. "Florida is sort
of . . . mindless," she says finally.

This sounds good to me. I'm tired of thinking. My
brain turns damp and humid at the idea of "developing a
relationship" with Dad. I could get worked up against
the whole idea of it, except my elbow is touching Adele
Parker's. She asks what Dad does for a living. I realize
I don't know. "Maybe nothing."

"There's a lot of that down there," she says.

I like the cool Spence-ness of her. Since we're the same
age, she's not playing a role, even though she was born
knowing how. She's over the Spence hump: the moving
in a pack, the passing of the joint from some parent's
private stock, the "sleeping over" at marathon slumber
parties in somebody's penthouse on Park. And she's not
pushing anything with the Radcliffe-Mount Holyoke-
Bennington plans for after next year.

Her past flashes before me. She's grown up in Indian
Trail shoes, an ocean of shampoo, an atom of makeup;
her toenails have never known kinky color. She's never
been near a Barbie doll. For the first ten Christmases of
her life she was taken to Lincoln Center for *The Nut-*
cracker. She's never danced with another girl, not even
in junior high. She studies without stereo. I wonder if
she'd talk to me if we weren't strapped into adjoining
seats, but this is the kind of thing I'm trying to stop
wondering about. So I do.

[98]

Incredibly, we're touching down at Miami International. Too soon, too soon. We're taxiing now, looking for the gate, lumbering earthbound. Nub's probably standing up in his case, staring with amazement at the floor. The New York types are already on their feet, snapping shut attaché cases like castanets. "I'm going to make a run for it," Adele whispers. I try to get my legs out of her way, but the backs of her knees brush against me.

"I want to make it look good for Byron." She gives him a great, lady-spy look; then she's up the aisle, weaving past the businessmen, vanished in a disciplined cloud of bright hair. But I've managed to get her Miami address, and I've written it with an unsteady hand on the boarding pass.

Byron and I let the aisle clear before hustling Nub out. "We better not mention helping a stowaway to Dad," Byron says. "We better be careful what we say to a stranger."

EIGHT

He—Dad—is standing at the gate, at this our third meeting of the season. But this one has got to take. He's wearing a pair of white tennis shorts, a denim shirt with turned-back cuffs, white socks with blue bands, and Adidas sneakers. It could be worse: gold chains around the neck, flowered body shirt open to navel, polyester leisure suit, patent-leather loafers.

The airport is Someplace Else, exotic. Dusky Latin beauties trip along in heelless platforms, heading for flights to Ecuador, Nicaragua, Caracas, swinging Saks and Neiman-Marcus shopping bags, wearing oversize purple-lens shades. All rose-in-the-teeth types. The p.a. announcements are in Spanish and English. We are off our turf.

It's high noon, or at least it feels like it. The sun falls in vertical lines, even indoors. We shake hands with Dad. Byron's sober as a judge. "Judge" may be the word. Dad gives Nub a slight double take. But then what's one more—or less?

We drive out of the airport tangle past the first palm trees shaking in the flight patterns. WELCOME TO MIAMI AMERICA'S FUN-IN-THE-SUN PLAYGROUND, followed by another billboard, splashier: BURIAL FOR TWO ONLY $295 COMPLETE. Dad has maybe the first Datsun ever introduced into this country. Bright orange, Simonized over banged-up bodywork. It isn't air-conditioned. You could bake brownies on the seat.

We thread through narrow streets, past little white cottages with electric-blue trim and religious grottos in swept yards. Very foreign. Tidy but temporary. Then we're on a sort of country road angling through the city. Enormous trees—banyans, Dad says—a dozen trunks snaked around each other and big umbrellas of foliage above, dappling the road. A bike path runs parallel, with cyclists wearing only sweatbands and shorts, pumping along beside us.

We pass a sign, COCONUT GROVE BUSINESS DISTRICT, and follow the arrow. Who is he? Lush? Lecher? Playboy? Pauper? All the above? The bonding on the Datsun's windshield is loose, breaking down the light into its component, rainbow parts. It hits Dad's legs. Tennis player? Swimmer? Over-the-hill jock? Who knows?

Wait and see. Plenty of time. An entire summer in a place with no other season.

We swing off the road, through the gate of a wall holding back jungle. Wind in through a track behind a big, balconied, raw-new apartment building, chalk-white above the green. The road wanders, years older than the building, through the bottom of an aquarium. Behind the high rise, hidden away, a damp little bungalow squats in sand and high grass. Practically an Aztec ruin compared to the new high rise. Porous coral stone walls, fading pink under a low tile roof. "This is it," Dad says, pulling into what yard there is. A solitary vine grows up over the entire house. I hear Byron thinking: is there room for all of us here? I wonder myself. It looks like a single unit cast adrift from a motel.

Dad cuts the engine. "The high rise is a condominium —the Barnacle Cove. I manage it. The house comes with the job."

Okay, we know a little bit more now. He's got a job, whatever it is. Janitor? Agent? Doorman? Window washer? What does a manager manage? I manage to get out of the car. My shirt peels away from the vinyl seat.

Dad reaches for Nub's case, but Byron has shifted it around to his good hand. "I can do it," Byron says. Good for you, kid. Show him. Show him what?

There are two bedrooms—cells side by side, and Byron and I get them. Dad's going to sleep on a couch in the living room, which is also a kitchen, divided by a counter. It's all musty but orderly. Clean clothes from a laundromat on the counter top, white socks folded in on themselves, army style. Byron and I get our room assignments. The situation's weird and familiar somehow, like the first day of the first year at camp. A summer of strangeness without idea number one about how to deal with it.

Byron has a small case for his clothes. I see Dad unzip the top and hand it to him in a single gesture. He knows one-handed Byron would labor over that zipper. But Dad doesn't try to do more. A nice touch, I have to admit.

I hang around in my eight-by-eight room as long as I can. There isn't that much to unpack. I commit the room to memory, wishing it was a memory: single bed, chest of drawers, tile floor brittle but not gritty, Venetian blinds but no curtains. There's something very boot camp about this barracks, but I have no complaints coming. I should kiss the walls because I have it to myself.

I'm standing around in this room that looks swept clean of clues. But clues to what I don't know. And I'm thinking hard now. When I open the door between me and Dad, it's confrontation time. I don't know how it's going to go or how it's going to affect Byron. But we're just one warped plywood door away from a general clearing of the air. We are going to lay some ground rules.

I get around to opening the door. Nub's out of his

case and deeply asleep on the refrigerator top, spindly haunches drawn up in a hump. He's shaped like a 1939 Plymouth. Dad's shucked off his shirt and has on a pair of faded red trunks. "I thought we'd have a swim."

So much for the big scene. I go back in my room, pull open the drawer where I think I've stashed my trunks. There's a totally different swimsuit in this drawer, female style: a bikini and bra for some unknown, trim little figure. Bright flowers on a black background make an expressive little two ounces of nothingness. A clue unswept away. Or maybe not. Maybe this particular item is left here because nobody's erased any clues. Maybe we're all open and up-front here. Or would be if anybody could say anything. I open the right drawer and find my trunks and pull them on.

The pool belongs to the high rise, at the far end of a path of crushed white shells. All around are bulldozer scars where they've carved all this instant luxury out of a swamp. Nobody's at the pool. The Barnacle Cove's inhabitants, if any, are sealed behind brown-toned windows in their air conditioning.

There's a kiddies' wading pool across the patio, with a very tame slide slanting into it. Byron walks past it without a look.

I'm ready to dive in from the side of the big pool when I remember Byron can't go in the water with that sling, much less swim. Dad, of course, hasn't given this a thought. He's probably one of those everybody-in-the-pool types. He's busy making a pile of towels, a shirt and pants of his own, and a beeper—those things doctors wear on their belts that signal them to call the hospital. Except he's no doctor. I see he's letting Byron find his own way to the far end of the pool. It's practically Olympic size, though kidney-shaped, this being Florida. Wide steps go right down to the bottom at the shallow end. Byron steps out of his thongs, hops once on the frying concrete, and walks down the steps into the water. He moves experimentally. The green water climbs up his white body to within inches of his arm. He backsteps and sits down on the steps in the water with the sling just grazing the surface. He looks around at the water with little eyes slitted against the glare. A jet-black beach ball floats toward him. He bounces it out of the pool with one foot. He's improvising a game that will keep the ball in play within range of his foot. One of his moods of complete calm is settling over him.

I dive from the side into the deep end and fight down until I can slap my hand on the bottom. The water's too warm at the top, but below it's like a night in the mountains. I think seriously about living the entire summer underwater, without benefit of scuba.

When I surface, Dad's sitting on the edge halfway down the pool, with his legs in the water. We're not going to be treated to any trick dives off the high board. I swim myself into a half-daze. Incredibly, the sun's actually beginning to shift off dead center. It must be five o'clock. Maybe not. Maybe only four. Dammit, at this rate we'll all be old men before this summer's over.

Finally I pull up out of the water a calculated distance from Dad. His beeper starts sending out a signal. He gets up, grabs a towel to dry his legs, and pulls on his clothes. "Probably somebody to look at an apartment," he says, nodding toward the building. "Want to come along?"

I shake my head, and he starts off, jamming his feet into his tennis shoes as he goes. At the end of the pool, Byron has the beach-ball game down to a science, almost juggling it out of the water with his feet. He's getting the maximum mileage out of this pool without even getting half wet. He's settling in.

"You want to eat in or out?" It's actually evening, practically, and we're all back from the pool and showered. Nub's been taken out once to do his business in the sand outside the door. The biggest litter box he's ever seen. He's scratched around and dampened Florida in twelve separate places and stared yellow-eyed up into trees full of invisible rustling birds.

"I don't care," I say. It sounds sullen, but I can't help that. I don't care whether we eat in or out or if.

"Out then," Dad says. "I'll have plenty of time to treat you to my cooking. You don't cook?"

Somehow in a room this small I can't look him in the eye, even when I'd like to be staring him down. "I'll try."

"No, it's okay," he says. "We'll manage."

I wonder.

We walk out into the evening and around the Datsun. We're going on foot. It turns out that the jungle plot Dad lives in is set down in a small town, Coconut Grove, which is somehow marooned in Miami. And no place exactly relates to any place else. Furthermore, Coconut Grove is not your basic small town. Though it's half shut down in the summer, it's a fairly funky place. Beads, beards, beach bums, the Hamptons or Greenwich Village ten years behind. They're still into macrame.

We stroll past a somewhat chi-chi place called the Coco Plum, jammed with people eating crepes in white iron chairs. I learn later that there are only two other bona fide places to eat in the Grove. Lum's, a routine short-order spot disguised behind driftwood siding, and a place with no name. We go into the no-name place, which is not trying to be anything but a diner with mimeographed menus. It's not doing a great business.

We settle into a booth, and a waitress who's having to handle the counter and the booths comes over. "Hey, Howard," she says. I don't see the look he gives her because this place also turns out to be too small for looking Dad in the eye.

"Hi, Marietta," he says. "These are the boys." Not

exactly an introduction. Evidently none needed. We're easing into something very casually. This Marietta seems to be in the picture already. I throw caution to the winds and give her a look.

Her eyes would be too big except they're violet blue with real lashes. The rest of her face curves in sharply from strong cheekbones that balance the eyes. I should be a painter and go into portrait work. Her face is heart-shaped, and her hair is like a black cap, twirled just a little at the ends. She's maybe twenty-four, twenty-five. I get that bell-ringing sensation that Interesting Older Women give me. Distant bells.

Even in her first three syllables there's a kind of Dolly Parton drawl in her voice, soothing-southern. She's not wearing an emblem T-shirt or a waitress uniform, just a middle-of-the-road white blouse over an old-fashioned apron banded around her only-big-as-a-minute waist. She's too thin, except it works. Her stock rises even higher on my private board when she gives Byron a fringed look without ruffling his hair or asking about the sling. Maybe she knows about the sling.

"What's for supper?" Dad says.

"These boys ever had grits up Nawth?" Marietta asks. I can feel her look fall on me, but I'm fooling with the salt and peppers.

"That's breakfast, Marietta," Dad says.

"Breakfast all day," she says, "our motto. You come back in the morning, sport. For grits." She taps my elbow with one finger. Being called sport is not my favorite thing, but she can make anything sound good.

There's more conversation. I think Dad must be starved for it. He banters back and forth with Marietta, all very familiar. So he comes here to eat a lot. So would I. She leaves without ever getting around to taking our orders. And comes back with an entire church supper. A Corning Ware dish of baked beans, ham-and-cheese sandwiches piled on a plate, French fries in a basket, coleslaw, pea salad, bright red Jell-O. Byron watches all the food covering up the entire table. This is not Almah's style.

"You cleaning out the refrigerator, Marietta?" Dad says.

"Bawd of Health," she answers, mysteriously.

Byron checks with me to see where to begin this orgy. I'm lost. There must be a balanced diet in here somewhere if we can find it. The red Jell-O comes in parfait glasses. "Is this dessert?" he asks, giving me owl eyes and pointing straight down at it.

"There's three slabs of raisin pie set back," Marietta tells him, though I think she's looking at Dad. "That'll be dessert so save room for it, you hear?" She tucks a paper napkin into the neck of his shirt and goes back to the counter. I half expect her to slide into the booth with us.

"Home cooking," Dad says, and starts passing dishes.

Marietta reappears when we've picked the table nearly

clean. "What's the matter with my baked beans?" she wonders aloud. "There's a dab left." We rub our stomachs, roll our eyes, show appreciation with body language. "Well, I guess I'll have to tote it home in a paper sack," says this gaunt, gorgeous girl, mother-henning us. "Save your forks for the pie."

Byron looks at his fork in disbelief. We've never eaten this well out of Almah's kitchen, but there's always been a separate fork for dessert. But we're far from Almah. We are, as I say, off our turf.

I'm lying awake in a strange bed, thinking I haven't been asleep, though I have. No air conditioner in the only window, but the muggy dampness from the jungle outside passes for coolness. The trees are loud with country sounds. Subtropical birds complain bitterly. Insects sing grand opera. All kinds of rhythms going in the undergrowth. These are sending me to sleep and then bringing me back. I haven't counted on the night being twice as long as the day.

Then I hear inside sounds breaking away from the din outside. A little muttering of talk, high and low. A shuffle of feet somewhere not immediately outside my door. Water running. The house is alive, and suddenly I know we're all three awake, but I'm the only one alone. I can hear Dad's voice rumbling and Byron's answering.

I'm out of bed like a flash, grabbing around in the dark for my jockey shorts. The tile adheres to my feet.

I crack my door, but the living room's dark. There's light past the refrigerator, outlining Nub's head and cocked ears up near the ceiling. The bathroom's just beyond the refrigerator.

I walk toward it. Byron's sitting on the edge of the bathtub. Dad's sitting beside him in a T-shirt and pajama pants. He's got his arm around Byron, whose breath is catching in his throat like he's in pain. He's rocking back and forth on the tub.

"What's wrong?" I say to him, stepping out of the dark.

"Nothing," Byron says, looking up quick, clenching the little knobs of his knees together. He's wearing only shorts, and his skin's already a little pinkish from the sun.

"His collarbone's hurting him," Dad says.

Oh, great, just when we get to the end of the world. "Is it bad?" I say to Byron, wanting to move in closer, except Dad's already there.

"Naw," Byron says, still rocking.

"I gave him aspirin," Dad says. "It'll take effect in a little while." I wonder how long they've been sitting there. "I heard him walking around in his room. It's going to be all right." He says this last part to Byron, and his arm's still around the slinged shoulder, lightly cradling it.

Feeling useless, I say, "I wonder why it's flaring up now."

"It always aches at night," Byron says. "It's like a bad

headache, only it's here." He dips his chin down to his right shoulder. "It'll get better," he says, encouraging me. "But the doctor said I'll probably always be able to tell when it's going to rain."

This interesting item gets past me. "You mean you've been hurting before and didn't tell me?"

I'm suddenly back in Brooklyn Heights, going over nights past and beyond recall. And I see myself crapped out in my bed at the back of the house and Byron sitting on the edge of his bed at the far end of the hall, rocking back and forth because his splintered collarbone is giving him fits. He doesn't want to wake me up. And I have not given him one thought.

Then I practically go crazy. Dad's sheet-plastic bathroom walls wobble and flicker.

"Get away from him," I say, starting quiet. "I'll take care of him." Then the floodgates really open up in my head. I'm cutting loose, louder and louder. "Damn you, get away from him!"

Dad isn't looking at me, but he isn't moving.

"Don't, Jim." Byron's shaking his head. "Don't say bad things."

But I'm chock-full of bad things, eight years' worth. "You! I'm talking to you!" Dad knows who I'm talking to. I'm leveling at him over Byron's head. "You leave him alone. You gave up your rights to him. You walked out before you ever heard his voice. You think you can make that up now with a couple stinking aspirin? God damn you!"

[112]

Byron's whimpering now, ducking his head, running his fist across his eyes. But the tears are falling anyway. And Dad's arm stays around him, shielding him. And still I'm not through with all I've got to say. I haven't even gotten a good start.

NINE

I wake up in the morning, already remembering I'm at a tremendous disadvantage.

Nothing I've said—yelled—the night before has provoked my dad to deck me. So I've got something coming. He just let me run down, then walked past me like an old man. Not my old man. Just any old man. And I was left with Byron, who was finally coming apart. But these aren't the healing tears I've been waiting for from him since Mom died. I did this to him. I've pulled the rug out from under Byron.

Even getting out of bed takes a giant act of will. I have no idea what time it is. I've already stuffed my Seiko in a drawer. In a summer this long I've decided not to divide time artificially. This is not slated to be a season when you need a sweep hand. I get dressed and notice my hand reaching for the doorknob. I figure Byron's sleeping it off. I hope Dad's out of the house.

He's sitting at the counter in front of an empty cup.

"Why don't you go on up to Marietta's for breakfast," he says. I miss hearing that this isn't a question.

"I'll get something here." Though I don't have a leg to stand on, I'm walking toward the refrigerator.

"No," he says, "I'd just as soon you get out of the house for a while. I don't feel like facing you this morning."

I know the feeling. It's mutual. So I walk out into the morning, up toward the street. Coconut Grove is a late riser. A few cars out, fewer pedestrians. A panhandler type wrapped in a winter overcoat is just stirring on a park bench—Skid Row with palm trees. The bum has a great tan.

I drift around the streets, past the Dade Cycle Shop and the Blue Water Marine Supplies. Up and down past little shops behind fake Spanish arcades. A bookshop, the Grove Book Worm, is doing some early business. "We Have *The New York Times*," says a sign in the window, enough to trigger twenty-four hours of pent-up homesickness.

Lum's is advertising a ninety-nine-cent breakfast, and I'm tempted by a completely anonymous place. But I'm heading for Marietta's, and my feet know the way.

[115]

"Hey, Jim," she calls out before I'm in the door. She's got four or five at the counter and a boothful. Her grits rush hour. I start for the counter, then veer off to a booth. I'm settled into it before I remember she's been introduced to me, but I haven't been introduced to her. Funny how she can pull my name out of a hat, so to speak.

I'm scowling at the breakfast menu, which is smudged beyond belief and plastic coated. She comes up with a coffeepot, slips the menu out of my hand and replaces it behind the sugar bowl. "Gonna be hotter than a scalded dog today if it keeps up," she says.

I manage to glance up at her, at least as far as her blouse. Her breasts are small and apple-firm. "Get a smile out of you?" Her hand's on her hip, and I can see she's waiting me out. I work up a smile that feels cracked in three places.

"Well, for Pete's sake," she says. "I wisht I hadn't asked." She's gone then, but I know her routine. The menu's a mere formality. The coffee, while good, is a big mistake. It's clearing my mind. I'm thinking at the top of my lungs.

Marietta's back, scooting a platter in front of me. "We fry in butter," she says. "It makes that little difference." Why is it everything this straightforward girl-woman says sounds vaguely mysterious? It can't be the accent alone, which isn't that thick unless she wants it to be.

At one end of the platter we have a slice of ham, four

[116]

strips of bacon, and numerous sausage links. At the other, scrambled eggs in a pile. Between is a vast mass of gray-white granular stuff. Grits. I eye them a moment too long.

"You salt 'em and you butter 'em," Marietta says. "Orange, grapefruit, or for the desperate, prune."

"Excuse me?"

"What kind of juice you like? I know Yankees don't have grits, but you mean to tell me you-all don't have juice?"

"Orange, please. Marietta."

She's off toward the kitchen then, and I think she's humming, "Come to the Florida orange juice tree. . . ." She's as corny as *Hee Haw*, and I'd like to rest a hand on her breast. And she's going to cheer me up if it kills us both. I'm wondering why I can't play my role in this. Then I remember I have my reasons.

I start stuffed and then eat with a growing appetite on both sides of the grits. Then, giving in, I slide the fork into the grits, which I've salted and buttered up a storm. They aren't bad. I don't want any more. But they aren't bad.

Somewhere from a distance I have the sensation that someone is checking on my grits-eating.

She comes back as I'm contemplating the naked, steak-size platter. And she slides into the booth. "Hey, Marietta!" comes a voice from the counter. "How about some more coffee?"

"Just go around back there and hep yourself," she

[117]

calls over, "and add it to your bill." This causes general merriment.

"Well, how you like Florida?" she says, smoothing out an invisible place mat before her. I look at her hands, her fingers thin between the joints. No rings. No ties? I think again of executing great paintings.

"You tell me," I say. "I'm a stranger here."

"It's heaven," she says, waving one of those hands around the totally tacky diner. She means it. "Where I come from, Florida is where you go when you die if you been good. I came on down early." For some reason I'm just barely following her train of thought.

"Listen"—she taps the table with one finger—"you come from Dothan, Alabama, and anything south of five miles north of Mobile looks like heaven on this earth to you. I remember the first time I saw the Gulf—around Panama City, Pensacola—Red-Neck Riviera, they call it—I thought: one day I'm going to keep a-going south until there is no more. And here I am. Miami's about as far south as you can go. Except for the Keys, and they've about got them ruint with them dope runners and what-all. Key West is a mess with all those weirdos from—"

"Up north?" I offer.

"All over," Marietta says, ladylike. I decide I have a natural affinity for people who don't look you in the eye. Her thumbnail traces invisible curving patterns on the table top, and her eyes follow, maybe seeing the patterns. I'm looking her in the lashes. Yet I know she's been studying me. And I like it, whatever her reasons are, even if she's seeing Dad in me.

"You'll see," she says. "You'll take to it down here. And your little brother will too. They got the Seaquarium and Planet Ocean and the Parrot Jungle and the Venetian Pool and dog races at Flagler. A snake farm, too, if that's your idea of a good time, and jai alai and the Goodyear blimp. I don't know anything about New York City, but oh, Lord, they got everything here a person could want."

"And what happens after a person's made the rounds of all the sights?"

"Oh well shoot, that'd take you a month of Sundays. I left out Walt Disney World up at Orlando, and Palm Beach, where Mrs. Got-Rocks lives, and—"

This is not the last time I hear of Mrs. Got-Rocks from Marietta. Mrs. Got-Rocks is her own private myth —the lady born with silver spoons sticking out all over her, who lolls around all day in Marietta's shifting idea of luxury: in a mirror-chrome bed dividing a mango with a solid-gold knife, wrapped in endangered species for a night at the opera, which is an art form Marietta takes on faith. To Marietta the world's an endless round of novelties, and she takes an innocent pleasure in the doings of the rich, forming and reforming them to fit into little pigeon holes in her mind. She's the most naïve person I've ever met, and full of wisdom. ". . . But of

[119]

course it's the people who make a place," she's telling me, and her eyes want me to believe it.

"And what kind of people live in Miami?"

"Oh—people getting away, starting over. . . ."

"Hiding out?"

"Them too."

As far as I'm concerned, the conversation's skating near Dad. I'd like to ask her point-blank if she's having a thing with him. The need to know leaves as fast as it came. I'd settle for a return to Mrs. Got-Rocks, but Marietta says, "Where's your daddy this morning?" spoiling everything.

"Sitting home in front of a coffee cup."

"He's a bear with hangnails in the morning," she says. "Like you."

"I don't know much about him. And I'm not too interested in being like him."

She gives me a look, runs a sweeping hand over the table top. "That's right," she says. "Be yourself. It's the best policy." She's pitched this bold pronouncement of mine back in my lap. I figure I've made her mad, since she gets up and walks away. But she's back with my check, hands it over with a perky flourish. It totals out at seventy-five cents. "Are you kidding, Marietta? This is ridiculous!"

"Got to undersell Lum's," she says, giving me a shrewd look, deadpan, mock-serious.

"At these prices you won't last the week," I tell her, meaning it.

[120]

"At these prices everybody in Miami should be eating here by the end of the week."

"Do you own this place, Marietta?"

"Shoot, no. If I did, I'd sell it. See you, sport." She drifts back to the counter, and I watch her go, the apron bow in the small of her back, the stockings glossing her slender legs above the bloblike waitress shoes. And I know she knows I'm watching her. She's been watched before.

I take my time heading back to the house. But he's sitting where I left him. He may not have moved. Not noticing the milky cornflakes bowl and the drained juice glass, I figure Byron's still asleep.

I guess Dad's dressed for work. His beeper's on his belt. He's wearing a spotless, unironed golf shirt. His face still has an early-morning look, what I see of it. I don't want to study that face too closely. I might see somebody I know in it. "Let's get it going," he says.

I decide to stand. Is it an item of interest to either of us that I'm taller than he is, marginally? "If you want an apology for last night—" I begin.

"I don't," he cuts in. "Apologies work between friends, occasionally between sons and fathers—nothing that covers our situation."

He's taking a fairly aggressive line, I see. I still have unspent ammunition from the night before, but the old steam just isn't there. To weasel out, I nod my head toward Byron's door, using this as an excuse not to talk. Or to be talked into anything.

[121]

"He's up and out," Dad says. "He wanted to go back to the pool."

Naturally, I picture Byron fallen into the deep end and floating there dead and bloated, sling out of the water like a broken mast.

"He'll be all right," Dad's saying. "If you give it some thought, you'd rather see him on his own than with me."

If I keep quiet, it means I'm giving this some thought, but I can't think of anything to say.

"He's out of earshot now, and if you want to pick up where you left off and light into me, go ahead. If that's your seventeen-year-old idea of clearing the air, cut loose."

He places five fingertips against five fingertips and stares into the cage this makes of his hands. He's waiting. "That's too easy," I say.

"Meaning?"

"Meaning if I mouth off for ten minutes, that wipes your slate clean for eight years. Screw that." I jam my hands in my hip pockets and give him my back.

"Great," he says. "Then maybe we can talk some sense." He actually clears his throat, which seems to declare the negotiations open. In order not to be drawn in backwards, I turn around.

"You're raising your little brother, right?"

"Who else is going to do it?"

"Okay. I've dealt myself out a long time ago, and your mother—"

"Hold it right there." I jab a finger at him, try to make him look down the barrel of it. "Leave her out of this. Completely. I don't want to hear one word about Mom from you now or ever." I see it, or I think I do. He's comparing his cutting out on us with her doing the same thing. I'm not taking this from him, even to use it against him.

"A rule," he says. "I can abide by that. You're raising Byron, but you're under my roof. This is one thing we can't blame each other for. It doesn't take much imagination to see this was your grandmother's idea. If I'd made a claim on you both, she'd have thrown me out of her house. But once it was her idea to send you two down here, then it was fine. All I want to know is, where do we go from here?"

"How about back to the airport?" I say. I have his forehead in my sights. With a bit more effort, I could be looking him in the eye.

"Your grandmother's gone to a hotel in the Poconos for the summer with that friend of hers from Joralemon Street."

Typical of Grandmother not to tell me her own plans. Typical of me not to ask.

"I doubt if she'd put up with the idea of you two rattling around alone in her house all summer."

How true. And what a great idea.

"So we're stuck with each other," he says, "and I want some kind of idea how we're going to make it to

Labor Day. If you don't tell me, then the ball's in my
court, which I assume is the last thing you want.

"To take a recent example, when Byron—or you—
gets sick in the night, I want to know whether to get up
and look after you or pretend to sleep through it. As
you'll be quick to point out, I've got no history of dealing with kids your ages. So you tell me, and I'll see if I
can live with it."

The question's academic. After the stink I've caused,
Byron wouldn't dare let his pain show again.

This guy, Dad, should have been a lawyer. He's
talked me into a corner, and it doesn't even have that
prepared-speech sound. If we've got any common
ground, it's the impossible dream of getting through this
summer in three pieces. To shut him up, I try to build
on that. "I think we can make it," I say. Actually I mumble it, but he hears. "I don't have any master plan. Let's
take it easy and . . . do what comes naturally. Let's
call it a truce."

There are holes in this, but he ought to be satisfied
with it.

"I don't know what truce means," he says, "but fine."
He holds up his hand, one stubby finger in the air. "Byron's the unknown quantity in this, though."

"Not to me he isn't."

"At his age they change fast. Every day. In a lot of
ways this summer's going to be longer for him than for
us. At my age three months is nothing—even these. At
your age—well, skip that."

[124]

"Better not," I say. "What's three months at my age?"

"You want it to be nothing because you'd like to think you're already grown up."

"And how do you work that out?"

He raises his eyebrows slightly. "Because I was seventeen once."

I want to say: and you're probably still seventeen behind that sagging exterior. But I let it pass, wanting to get back to Byron, which he does.

"But he's going to be doing some growing up this summer. He's not going to be the same at the end of it, and in a few ways that have nothing to do with you or me."

I don't see what he's getting at. Later, I see, but not then. "Don't worry about Byron. He's very mature for his age."

"No, he's not, as a matter of fact," says Dad, the big authority, the self-anointed Dr. Spock. "He's a typical kid, a little quiet, a little bewildered by the death . . . by a death in the family. And he's not a minute ahead of himself."

I ooze with sarcasm: "Of course you'd know with your vast experience of him."

"Could it be I can see him better than you can because I'm not breathing down his neck?"

"No."

"Okay. So that makes two people we're not going to be able to talk about. My ex-wife and your gifted brother. And as far as I'm concerned, you can add your

[125]

grandmother to that list. It looks like it's just you and me, by your own decree. Anything else to cover before the truce?"

"Yeah, a couple of things." One of them's my karate chop, and I find I've had it ready all along. But it'll keep a minute longer; I save it for last. "Let's not have any talk about the so-called good old days."

"Which particular good old days?"

"I'm not surprised you've forgotten," I say. "I'm talking about the time before you walked out on us when Byron was four months old."

He eyes me. Runs his tongue around his lower lip. "Well, I suppose from your viewpoint that's going pretty far back in history to build anything on."

"I'm not interested in building anything. And let's not have any talk about the future either. No point getting Byron geared up about something he's not going to have. You might as well know one thing right now. After this summer's over, I'm going to see to it that he forgets we were ever down here."

Some of this speech comes to me as I say it. Dad sits there. He seems to listen even after I've run down.

"All right. More rules," he says. "Looks like we're stuck with the present. Remember, you called the shot. Anything else?"

"Yeah." Here it comes, the grand finale. "You sleeping with Marietta?"

He doesn't bat an eye. Here's his second opportunity

to give me a fat lip—come on, creep, let me have it. He just lets the question hang for a long moment. "You're going to be easier than I thought," he says. "You've got no subtlety. Now."

He drops his fist on the counter top like a gavel. "The truce begins."

TEN

It's July somehow, and for maximum mobility and to work off steam Byron and I rent bikes from a place with a poem in the window:

> Buy a car nevermore
> Remember: Ten on the sprocket
> Not four on the floor

Byron's out of his sling, checked over by a local doctor looking strangely like a surfer: striped tank top under his white medical coat, a Moped in his reserved parking space.

We've pinked, peeled, and are working up authentic tans. Biking browns the shoulders, clears the head, or so

I think before I come unstuck. We range all over the Grove and the territory across the Dixie Highway. And this is where I'm nearly totaled one innocent-seeming afternoon by something that creeps up behind me. Not traffic, something else.

Byron's pumping along ahead of me, and I'm letting him outdistance me when we ride past a kind of vacant-lot park. There's a woman in the middle of the park, braving the heat and feeding two little kids a picnic on a couple of beach towels.

The light flickers through the palms, and I catch a glimpse of the woman in profile, do a double take, and nearly end up under my bike. My foot grazes the spokes. Even the little kids are familiar. One's staggering around on the towels; the other one's older, sitting still.

I squeeze the hand brakes, give the front-wheel brake extra grip. The rear wheel leaves the ground, and I nearly lunge forward over the handlebars. I've got to stop and make contact with these people. The bike wheels scrape the curb, and Byron's a half block ahead of me.

I plant a foot on the curb. The woman's not really close. But still I know her. She's coaxing the smaller kid to eat something, and I know the gesture of her out-stretched hand and the way she balances back on her arm. I know that coaxing little movement, and I can almost hear her familiar voice.

I ease the bike flat, forget about Byron, who hasn't looked back. I'm across to the sidewalk and then a step or two into the park. The smaller kid sits down suddenly. His legs fly up. I hear the woman's familiar laugh. It wells up in her in that old way.

Then I stop dead, just short of her noticing me. And I turn around, sun-blind, toward the empty street. I thought this woman was my mom. Crazy? Yes. Did I forget she was dead? That she wasn't as young as this woman? That those kids can't be Byron and me because—how could they be unless I'm the dead one, looking on from some greater distance than I am.

The ground lurches, and I'm still blind. And shaking like a leaf. I can hear somebody crying in another room, except here we are—I am—outdoors on a day baked into silence.

What's the real meaning of this? My mom died weeks —actually months—ago. Am I just now noticing? Hell, is Mom some item, some gym shoe I've lost at the back of my locker and didn't even miss till I came across it again by accident?

I'm sweating and crying. What have I built up all these weeks that can come crashing down on my head this quick? I'm standing on the curb of a street with some unknown name in the middle of a city I haven't actually paid much attention to. And there's nobody in sight to use or blame or lean on or anything. And I want my mother back.

The tears are all over my face, and I don't wipe them away. My hands are still clenched from gripping. And they're hanging down at my sides. I feel ape-like, grunting with grief.

I come to in stages. The first half-sane thought I have is that I'm hurting because now I'll never know my mom. Not as a person. I don't even have the option of forgetting her. I'll only remember the role she played and won't let myself remember much of that. Not at what it's costing me right now. Finally I know what *loss* means. The power of the dead is that they leave you with the living. I'm staggering under this load of new knowledge, but what do I do with it?

"Do with what?" Byron says. He's doubled back. It looks like he's walked his bike the last few feet toward me. He's standing in the curve of the handlebars, and he's squinting up at my wet face, and his eyebrows are meeting.

I reach for believable lies. I've got a flat. The gears are shot. My sprocket's sprung. They've suckered me with a defective bike.

"Why are you crying?" Byron asks, and waits.

"I'm crying about Mom," I sort of whisper. I'm so far gone that I'm telling the truth. It sounds like a foreign language.

"Oh." Byron looks down at the wet tar that snakes all over the pavement. He's half embarrassed, half something else—respectful maybe. "It's okay," he says in a small, thin voice. "I do it myself."

I can't talk. I can't let him see me like this. But he's

seeing. Everything that seems to have worked for weeks grinds to a halt. The old systems are shutting down. But what I really can't do is talk around these tears. Still, I do. "I . . . really . . . need . . . you . . . a . . . lot," I say to him, wringing out every word.

"Well," he says, "sure."

I bury Mom again, less certain this time. I scatter her ashes over this subtropical street, straddle the bike, and wheel off behind Byron. He's standing up, riding the pedals slow and easy. The wind catches in his hair, and I ride in his wake and feel it.

In a few hours, days, I settle into life with my fellow survivors. We touch up the Truce, Dad and I, and we don't bog down over petty details. We iron out some with a couple of words and walk around the rest. A little goodwill seeps in, depending on the day. We're at our best with trivia. There's even an unwritten roster for taking showers so the one left with the cold water is alternated. Byron develops a thing for cold showers, one of the big guys.

Okay, so I fire a few more zingers at Dad occasionally. Quickies, hit-and-run, sniping out of a tree. One time he gets personal, testing the future clause of the truce. He asks me what I want to be when I grow up. Rephrases by asking what I want to be when I get out of college.

"Marriage counselor," I tell him. He turns the other cheek.

Mostly I'm at my worst when I figure he's about to take Byron over, which he never quite seems to be doing.

Or when I imagine Byron's drifting in his direction. After one of my verbal jabs—I forget which one—Dad says, "Take your best shot; I return fire." But he hasn't yet.

While we have our day-to-day rules, we aren't slaves to routine. The house looks pretty much the same, clean or not. If Byron should wet his bed, which he doesn't, nobody's going to rub his nose in it. We do laundry only when it's past counting. We bundle everything into a sheet: shirts, towels, sneakers, socks, jocks—and load the Datsun down for a trip to the laundromat.

"Boys' Town," Marietta calls our place.

"It needs a woman's touch," Dad remarks.

"Shoot, a woman wouldn't touch it," she replies.

I encounter Dad in unexpected places. Beside his sofa bed there's a night stand/end table/magazine rack. At the bottom of a pile of paperbacks I'm browsing through, I find a heavy hardback volume, expensive binding, gold-stamped.

His college yearbook. With a few minutes to myself, I leaf through this ancient document, unsticking its pages, and get caught up in it long before discovering Dad. There's a sort of 1950ish centerfold of beauty queens. Big, luscious, flat, black-and-white camera studies of re-touched ice princesses, all seeming to wear the same strapless black dress. I know I'm not going to come suddenly upon Mom because she didn't go to this college. So I look my fill: at Miss Beth Bartley, Delta Delta

[133]

Delta, Homecoming Queen; Miss Nancy LeDue, Alpha Omicron Pi, May Day Queen; Miss Elaine Lindstrom, Kappa Alpha Theta, Military Ball Queen; Miss Alberta Cahill, Delta Zeta, Queen of the Interfraternity Ball.

The light catches their teeth and pearls. They smile down the years, not aging. Which I guess is what year-books are for. I flash through the pages of this forgotten world, a weird middle-earth kind of place full of spindly halfbacks in action, cheerleaders doing splits, track stars with horn-rims taped to their temples, the rifle club posing around a gun rack. Wrist corsages, debate teams, the Methodist Student Movement, a couple in incredible formal attire in an arty doorway shot, only their lips touching. It's spooky, but I like it. Everybody's smiling.

I come across Dad in the senior pictures. He's midway down an alphabetical row, but I never actually spot him. They've all got convict haircuts and clip-on bow ties, and they're all postage-stamp size.

But they've included his credentials. I read through his entry, trying to decipher it: "Atwater, Howard: Van Wert, Ohio." I try to remember his hometown, my grandparents on that side, but all I come up with is a car trip we took when I was pre-school. All I remember is a front porch with a ceiling painted sky blue and a porch swing painted green and orange.

Business Administration major. Delta Upsilon, rush chairman. ROTC. Interfraternity Council. Fresh.

Soph. softball. Track mt. pep squad. Yearbook business manager. Alpha Phi Omega. Poly Sci Roundtable. Fresh. debate. Economics Club, treasurer.

It's all Greek to me, even the part that isn't. I give up on it, rebury the book under the paperbacks.

I don't snoop, but I notice Dad's got a closet full of well-seasoned fishing gear, fairly elaborate, evidently his chief possession. But he doesn't so much as tie a fly in our presence. He hasn't taken us fishing, or anyplace else. He's decided entertaining us is not the way to go. Besides, he's stuck close to his job, sold all but two apartments in the Barnacle Cove. He works on commission, so whatever he earns, it ebbs and flows, with dry spots. Becoming a sudden family man may have lighted fires under his initiative. Two more units to unload, and he'll move on to managing another building.

They'll level his house for the "subtropical gardens and sports complex" promised in the Barnacle Cove brochure. And he'll be living somewhere else. I like the impermanence of this; it rules out the future, enforces the truce. We're ships passing in the fog here; a couple of hoots from the horn, and good-bye Charlie. I lie awake nights picturing the bulldozers coming in through the wall, plowing the place under. Who's looking for roots?

We've fallen into a meal routine that cuts confrontations to the bone. I go by myself up to Marietta's for breakfast—the way to start a day. Lunch is a grab bag. Byron takes an apple and a banana and Nub into our jungle. I decide it isn't my mission to teach him nutrition and table manners. We're not running a finishing school here. Besides, we may make it till fall; then he'll have all the scheduling he can use.

I'd be hanging looser than I am except the truce gets a little—weighty. On details we're golden. But all the big issues are taboo. It's like living in a country with perfectly workable traffic laws and no constitution.

Let Dad start playing Papa with us, and I'll nail him. But as a completely reasonable roommate, he's getting on my nerves. I'm working up to the final summer scene, when we shake hands all around at the airport (the same day they bulldoze his house) and Byron and I walk away, forgetting to get his forwarding address.

But something's already crabbing this act that I'm rehearsing. I have this nagging feeling that Dad owes us— me—some explanations. But that would involve the past, which is strictly forbidden by the famous truce authored by yours truly.

As Dad says, we're stuck with the present. The chief advantage of this present is Marietta's regular dinner-partying. We all go up to the no-name diner for a late supper every night when her evening "rush" is over and she can sit with us.

She manages to make an event out of every meal. Most nights there's something idiotic to celebrate. She sits with Byron on his side of the booth, leaving Dad and me shoulder to shoulder. She maneuvers this, refers to Byron as "the boyfriend." When he's out of his sling and eating with both hands, the two of them are very big on body contact. He's practically in her lap. The whites of his eyes glow in his berry-brown face. He's putting on weight.

"Tonight," Marietta says at one of our feasts, "we're having something special." She's explaining a carrot cake with a single candle on it, unlit. The candle reappears for various occasions. "It's Jefferson Davis's birthday." Her violet eyes are putting us on.

"You check that out, Marietta?" I ask.

"No need," she says. And she and Dad finish in chorus: "Where I come from, every day is Jefferson Davis's birthday." Byron makes a mental note to find out who Jefferson Davis is.

"Three men," Marietta always says, her favorite saying, "equals no conversation." She builds on our truce silence, turns us into an audience, a family, whatever suits her occasion. She skates over our wordless surface effortlessly at first, executing figure eights. "When you're number nine in the Bethune family of Dothan, Alabama," she says (often), "you've heard a power of conversation in your time."

Then she returns to her favorite theme, the great day coming when she'll steal Mr. Got-Rocks away from Mrs. Got-Rocks and "be sassier than a born-again C.B.er."

"You know where I'll live when I'm the second Mrs. Got-Rocks?" She digs Byron in the ribs. He's all funny

[137]

bones, Marietta notes. He's developing a giggle, rising out of a gurgle. He's halfway to a belly laugh. It's only days away. We all know where Marietta will live in her fantasy future. Palm Beach she's never seen, so it's to be Coral Gables.

Byron and I have biked all over Coral Gables, which lies off Coconut Grove's flank, in a different world. An upper-crust suburb on a wrought-iron and pink plaster theme with banyan boulevards and mansions backed up to their own yacht canals.

"I'll be stretched out there on my shez long," Marietta predicts, "and even them little spaces between my toes will have the best tan you ever seen. And I'll have a butler to bring me out my lunch on a tray—a little butler, cute and Cuban. Your size." She elbows Byron.

But the real Marietta breaks through the dream. "And I'll have me a garden run right to the street with a border of quartz rocks. Spearmint for iced tea, garlic for salads, beefsteak tomatoes, mushmelons, pie squash, peanuts, roast nears." We all have to work on this last crop. It becomes roasting ears, which translates to corn on the cob.

"And"—she scans us—"I'm going to set aside one night a week out of my busy social schedule to entertain Yankees."

"We'll come," Byron says, totally taken in. The fact that we're all three in love with her, revolve around her, spurs her on to greater heights. We're her groupies.

But questions start to form in her eyes, even when her

[138]

own conversation is flowing free. Half our weirdness she puts down to Yankee ways. She sees we're not a family, not in her definition of the word. But she's too careful, too kind to ask us why not.

"My daddy," she says, pulling him out of the air, "whoooeee, you didn't want to put your tail across the line with him! Whuppin' for one was whuppin' for all. Had him an old Harley-Davidson and tore up the slab with it. But Sundays? We took up a whole pew, and my daddy'd boom out them hymns—'I Come to the Garden Alone' and 'Still, Still with Thee, When Purple Morning Breaketh'—till you couldn't hear yourself pray.

"Whoooeee," she says softly, shaking her head at me, not thinking about her daddy at all, wondering instead why I don't want one.

Sundays, her day off, she spends with us. Yes, the flowered bikini in my drawer is hers. She changes in my room. I don't care how long the bikini's been there. Now is what counts.

She never goes in the pool, hasn't been near water since she "was throwed in the branch" by her brother Peadell.

We don't believe anybody's named Peadell. "What's his real name, Marietta?"

"I told you—Peadell. You don't want to hear his nick-name."

She never tans. Well, only a little. When she sits with arms locking her knees, there's a curving wedge of paler skin just beneath her bikini. Her waist is child-small, and her legs are twice as long as they seem behind her apron. She could be the center fold-out for *Farmer's Almanac*. Following the sun that hardly touches her, she migrates around the pool. She never wears sunglasses, only an old straw gardening hat that must have come down from Dothan with her. The hat band features a repeating pattern of Confederate flags.

I unfurl a beach towel next to her. But I don't move every time she does. I'm cautious, never want to hear the term "puppy love" from her lips. When I pull myself out of the pool—Mark Spitz without medals—Dad has zeroed in. They're lying side by side on their stomachs, talking quietly into their towels. They slip back in time, shifty as teenagers, into a pre-summer pattern I don't think about.

I pretend to be bushed by the pool workout and lie with heaving chest on the concrete, watching them through squinting eyes. Blue veins in knots stand out on the backs of Dad's legs. Blue spiders on the ankles, branching blue road-map lines up over the calves, bulging blue worms massed behind his knees. He's too old for her. There are a thousand proofs of this. I have the entire thousand filed alphabetically, ready to pull. He's got a *daughter* who's not that much younger than Marietta, for Lord's sake. Funny how Lorraine never happens to come up in conversation. I earmark an item in my mental file. In Florida it's tough staying uptight, but I find ways.

Byron's become an expert back-floater. He drifts all

over the pool; deep end, shallow end, it's all the same to him. And he has the buoyancy of an inner tube. Also, he's developing a small potbelly. With skinny legs trailing, he floats at the surface like some cartoon jellyfish, deaf to the world behind the ear plugs I make him wear.

We've been looking for a reason to celebrate, dine out in style on Sunday night at the Coco Plum. It's near enough Bastille Day, though Marietta points out that they don't observe this in Dothan. And the Coco Plum specialty is crepes. So this decides it. We'll eat French pancakes to celebrate the French Revolution, and Marietta won't have to cook.

We gather up the towels. Byron staggers up the shallow-end steps, doing his Creature from the Black Lagoon number and popping out his plugs. We're halfway down the path when Marietta says to him, "Where's that ole cat you used to have?"

"Oh, he's gone," Byron says, twirling the underwater goggles he always brings to the pool and never wears.

Dad and I exchange glances, for the first time. We're both thinking the same thing, both trying to remember the last time we saw Nub. And can't. Subliminal Nub. Deadweight Nub. Moron-king of the refrigerator top, never caught up on his sleep. Memory's beginning to serve us. Nobody's kicked over Nub's milk bowl in a week. Nobody's found a fur ball, shimmering like a slug, in the sink. Nobody's discovered a flattened Tender Vittle stuck into a sneaker tread lately.

"What do you mean he's gone, Byron?" Dad puts a

hand out, reining him in. "Is he lost?"

Oh, great, he's lost his cat, and he's wrecked by it, and nobody noticed. Except, of course, Marietta.

"Why didn't you tell us?" Dad says. We have a crisis here. Marietta senses it even though house pets don't play a big role in her background. She has the universal look of somebody who's just put her foot in something.

"It's okay," Byron says, watching the goggles flashing around his finger. We're standing in the scaled-down jungle, and I picture Nub eaten by hyenas or smeared halfway to Homestead on the Dixie Highway. "He likes it outdoors," Byron's explaining in a let's-not-make-a-big-deal tone. "I'd take him out, and he really liked the woods. First he went up a tree after a bird and fell down. But up he went again. I never saw him get a bird, but he caught a mouse—or something like a mouse. When I saw, he'd eaten most of it. Maybe he didn't really catch it, but he ate it. There are raccoons out here, too, tame enough to come up for handouts. Nub would dance around them, not too close, but he liked them. He goes after lizards too—those little gray ones like dinosaurs."

Still, we stand on the path, wondering how to take this. "But you put out water and cat food for him, don't you?" I say. "You want him to come back? You don't want to lose him?"

Byron shakes his head, not needing this attention. "He's gone back to nature." His head swivels around at the trees. "He knows where water is out here. And he finds food. Cats are hunters," he explains. "Oh, one night he came back and jumped up on my windowsill and looked in. But when I came to the screen, he jumped off and went back in the woods. He was just checking on me. He's an animal. He likes being free."

The evening sort of falls apart. Why am I not glad to have Nub out from underfoot? Why is it I'd rather see Byron upset about this? The damn cat was the only thing that was ever completely his own. It wasn't some toy he'd gotten tired of and gradually outgrown. He'd actually trained Nub to be independent and then watched him creep away, back to nature.

Dad's looking rattled too, which I don't mind nearly so much.

Somehow the Coco Plum plan peters out. Marietta warms up canned chili, miraculously finds cayenne powder at the back of our kitchen cupboard. We eat in cereal bowls, with oyster crackers, sweet dills, divvy up some sherbert from the refrigerator's freezer. It's not an event, but we eat.

"I raised a piglet from a runt one time," Marietta says, re-inventing conversation. "Bottle-fed him up to where you wouldn't know him, and let him sleep in a drawer behind the stove. Course you know what happened when he come of age . . . ."

"Bacon," Byron says, scooping up a mittful of oyster crackers.

"Chitlins," Marietta adds.

"Ham sandwiches," Byron counters.

"Pickled pig's feet," Marietta concludes. They start giggling. Dad looks at me, raises his eyes, lifts his shoulders. Crisis past.

ELEVEN

This August morning starts out like a full-color production number. It's a day like no other days. The air's full of promises. Rain in the night and thunder in the bay. Water's still plopping off the palm leaves. My room's full of washed green shadows. It's almost cool.

I arise . . . aroused. I've been dreaming of Marietta or some fantastic country-and-Western starlet cast in her part. All night this figure from Central Casting has lounged dreamlike around the pool in Marietta's bikini. No, truth to tell, she's wearing less than that. The details return in the dawn. My number on the cold-shower roster has come up, and I'm a prime candidate.

I do a little barefoot soft-shoe shuffle to the shower,

throw my head back and gargle the ice water, expect great things from the stretching-ahead day. Then heat kitchen-kettle water for a careful shave to crown my mood. Stride from room to room buck-naked, feeling mildly X-rated.

Swimming laps is paying off. A hint of definition is peeking through under the pects. My shoulders don't quite fit into the shaving mirror. Maybe they never did, but today I notice. I lather up face, chin, overdoing it somewhat. I let the razor bite back my sideburns a half inch—the only serious shaving the razor has to do. Stare into the bubbly mirror in search of emerging image. Do things you do only in a mirror: flare a nostril, cock an eye, consider contrast between teeth and tan, create an actual part in my hair and slick it down like a hopeful Princeton freshman.

Lorraine's letter is discovered on the counter where I've been walking my bobbing manhood back and forth past it a half a dozen trips. The letter looks like it's crossed the Atlantic in a bottle. Forwarded from Brooklyn Heights, then from Grandmother at the Buck Hill Falls Inn, then on to me. Little does Lorraine know where her letter has finally washed up. Would she go all soft and mushy to learn Byron and Dad and I are reunited under a single roof? Only if she knew.

She's had her baby; this is the announcement. Nine and a half pounds, a mere nothing, given Lorraine's size. "The whole birth was easy as falling off a log," Lorraine writes. A terrible image, terrorizing even, if you take it literally. But then Lorraine was not a Creative Writing major. They've named it—him—Carl because he was born in Germany. Could be a lot worse: Rudolph. Siegfried. Wolfgang. Her writing sprawls all over the airmail page. Is she writing while lying down or falling off a log? No, this is the way she writes. X's for kisses to Byron and me and threats of baby pictures to follow.

I check the postmark. Lorraine's baby is already three weeks old. It's August already. The endless summer takes a lurch forward.

I stick the letter in my shirt pocket, under the alligator emblem. I haven't figured out a use for this letter, or why it should have one. I head up the path to the pool, meaning to inform Byron of his uncledom.

What I find is beeper-belted Dad giving him a poolside haircut. A light trim, but close contact. Byron's sitting on a deck chair footrest, draped in a beach towel. Dad's squatting down, snipping away. I let this pass in the name of economy. There's no point in throwing dollars away on such monotonies as haircuts. I let them off with a wave, and cut around past the grand entrance of the Barnacle Cove on the way to breakfast. The concrete canopy over the front drive is already developing liver spots from the damp. Dad had better unload the last unit in this turkey before it falls down.

Even downtown Coconut Grove is astir in this magic morning. There's a version of autumn in the air. The tourist season, wonderful warm make-a-buck winter, is only weeks away. The girl and the guy who run the Levis-n-Tops Shop are out on the sidewalk, up and down a ladder. They're fitting a new awning over the storefront. Midnight blue with white fringe and tassels between the scallops. Upgrading from funky to chic. The Grove's showing signs of making a giant leap forward from the Sixties to the Eighties. Time is still ticking, after all.

"What-a-ya think, man?" the guy says from the ladder top, running ringed hands over expensive blue canvas. I don't even know these people, but it's that kind of morning.

"Far out," I say, in his language. "Class."

"That's the look we're looking for," he says down to me. He can't keep his hands off the canvas. His bib overalls are tapered, flared, maybe lined with Hong Kong silk.

Marietta's counter is full. A solid row of work-shirt backs and elbows. She's working them like an assembly line. Sends me a high sign with the hand that isn't pouring coffee. I can wait. She's still last night's dream, fully clothed, aproned. On a paper napkin I do elementary math. When I'm, say, fifty-eight, she'll be sixty-four— sixty-five tops.

I'm not taking this computation to heart. I just like the mental gymnastics of telescoping time. I like the way our lives overlap with loose ends. If she and I were the same age, she'd be hundreds of miles away—wherever Dothan, Alabama, is—getting ready for senior year. Simmering in a summer of kudzu vines, picket fences, hollyhocks growing right up to the slab, cheerleader clinics, drag races. She'd be ringing up sweet milk and groceries at U-Tote-M with some pimply sacker slobbering all over her. Her added years draw us together, keeping the Real Me a dark secret from both of us. The real me being a male virgin wallflower in a nervous knot of fellow sufferers—put Kit Klein in there—sweating out a Van Cortlandt/Spence mixer.

The work shirts rise in a body—a chorus line of construction boots—give Marietta one last leer, pay up, poke the toothpick dispenser, and shuffle out.

"Street crew," she says, bearing down on me with her bottomless coffeepot. "They're patching the slab on Ingraham Road this morning. Keep me on the hop while they're here, but they all leave at once, thank the Lord."

"Good tippers?" I inquire.

"Better be," she says, suddenly shrewd again, "if they want lunch where they had breakfast."

"I like watching you work," I say. This has meaning. I like watching her *breathe*, and I don't just mean the rise and fall of those apple breasts behind the blouse. But working, she's—poetry in motion. I don't say all this, of course.

"Story of my life," Marietta says. "Men never tire of watching women work." But she gets my drift. When

[149]

she reaches across to pull the sugar bowl my way, I go for it at the same time. I'm working on my timing. My hand covers hers for a moment. She gives me one of her you're-quite-a-little-guy smiles. I'm very anxious to establish the distinction in her mind between me and Byron.

"Come on," she says. "Got something to show you."

"Can't." I shake my head. "I'm stuck in this booth."

She draws me out, her hand closing lightly over my wrist. We thread our way out back through the kitchen I've never seen, past the cook who never looks up—a hobbit in a paper cap, feeding potatoes into the jaws of a mechanical peeler. Outside the screen door is a back yard, sealed off from the street. In a sun-baked el between the diner and another shacklike structure Marietta's created a garden.

"I just putter out here when I get a chance," she says, downgrading the very place she's brought me out to see, to share.

It's not the Brooklyn Botanic Garden. It's . . . something else. The night rain has washed it clean, laid its dust. The whitewalls on the tires she's planted flower beds in are snow circles. She's lined a path with pop bottles, necks planted in the sandy soil. A double row of green and brown bottle bottoms puddle the sun, divide the path from straggly cactus, ferns, mostly plain dirt. Institutional-size fruit cocktail cans hold potted palms.

"Nothing I'm used to wants to grow down here," she

says. "I take no credit for the hibiscus." The hibiscus that grows all over Florida is in full yellow and red flower. It pays no attention to Marietta's plan, sprawls, tries to scale fences into parking lots. She's built a birdbath out of a piano stool topped with a blue enamel wash basin. It stands in a circle of crushed shells. Past it, against the woven-wire fence, the end of the garden, is her single piece of lawn furniture—a buckled bench that was once a diner booth. "Oh, one of these times I'll have me a real garden," she says. "White wrought-iron furniture. A shez long."

She's—we're—thinking of her Coral Gables garden, but this isn't the place to mention it. She nips a hibiscus blossom from the fence and plants it behind her ear. Red and classy against her black hair.

We sit on the bench, our legs in the sun. She rests her feet, props rubber heels in the white crushed shells. She's shared her garden with me, and I have just enough sense not to praise it. She'd put down my praise. I take her hand, and we lace fingers, rest them on the bench between us. I'm back in last night's dream without the big anxieties. We're brother and sister.

My mind goes to Lorraine's letter in my pocket, but the sun's stunning me. Still, I finally manage to break the spell. I'm at my best in ruining good things.

"Where do you live, Marietta?" I never even wondered before.

"Right here in this garden if I could," she murmurs.

Her eyes are shut against the sun. She stirs herself. "Right now I'm staying with a girl over on Seminole. We're two to a room, but she works nights."

"Sounds like close quarters."

"Believe it."

"Why don't you get out?"

"Might do that."

"When?" (Why am I pushing this? Maybe she can't *afford* her own place, for God's sake.)

"One of these times. End of the summer maybe," she says.

My mind skips a beat, followed by a flash of lightning. I'm not the only one sitting out the summer. She's waiting for Byron and me to leave, and then she'll move in with Dad. Or more likely—the bikini in the drawer—she'll be moving back in with him after this annoying interruption.

The Hardy Boy has reconstructed the crime. The puzzle piece falls into place beneath his unerring hand. After a bit of fiddling with the knobs, we have a clear picture here.

My hand, not too steady, goes up to Lorraine's letter. "I heard from my sister today," I hear myself saying. I've pulled my other hand away from Marietta's, needing both to open the envelope my sweat has sealed again. I wait, to hear if she knows I have a sister.

"She the one married to the soldier boy?" She knows.

"Air Force."

[152]

"How's she doing?" I offer her the letter, but she looks away. She doesn't read other people's mail.

"She's had a baby," I say. This is the big punch line. How come I didn't know?

Marietta sits up, draws in her feet, smiles the first real smile of the day—pure pleasure. "What she have?" Now she's almost capable of taking the letter off me.

"A boy."

"Nice!" She settles back into a reverie of talcum powder, Pampers, bassinets. I see her seeing a crib with a big blue bow on it and a dangling mobile. Small baby feet doing high kicks. Somehow this isn't the direction I want her going in.

"What'd your daddy have to say about that!" Her teeth catch her lower lip in a smile to coax all kinds of warm, sentimental tidbits from me.

"I didn't tell him."

Her smile fades, and there's puzzlement clouding her eyes. "Why not? You seen him yet this morning?"

"Saw him, but didn't tell him."

"You mean you didn't tell him he's a grandpa?" I can feel her wonder without looking.

"Yeah," I mumble. "Think about it. He's a grandpa."

She doesn't think about it. "He'll bust his buttons when he hears! You better scoot along home and tell him."

"Look, Marietta, I'm Jim, not Byron. I walk. Sometimes I jog. I've been known to break into a run. But I don't scoot."

[153]

"Well, whatever . . . ," she says, mystified, drawing back.

"The point is, Marietta, your—my dad is a grandfather. Get it?"

"I got it first time around."

I should hang it up right now. I should scoot along home. It's going to be all downhill from here on. But no, I haven't made my point. "Marietta, he's a grandfather. He's too old for—skip it. Forget it." Still I don't get up and scoot out of this found-art garden while the scooting's good.

Marietta's shifted on the bench, studying me. But I'm concentrating on my feet digging furrows in the white shells she's put there. "My daddy was thirty-nine first time he was a grandpa," she says softly.

"Yeah, well, your daddy wasn't making it with a chick half his age at the time, I assume." The morning flies apart in fragments. I have wasted it, in the worst meaning of that word.

The minute I've said it, I don't believe it myself. I want to reach up in the air and pull the words back, jam them back down my throat.

Marietta sits unmoving, her eyes on me. Then she reaches up and takes the hibiscus flower out of her hair, turns the stem in her hand. Puts it down on the bench and then brushes it off on the ground.

"Well, I ought to be getting back," she says. "I've stayed out here too long as it is." She starts to go.

"No, don't," I say. I'm sick to my stomach. I'm sick of myself.

"I don't take it personal," Marietta says, softer than before. "I know you're out to spite your daddy, not me. That's your burden, not mine." Then she's moving around the birdbath and along the bottle path, away from the snake in her Eden. The screen door snaps shut behind her.

I have to follow her. I have no place else to go. Crazy Louise, the beachcombing shopping-bag lady, is the only customer in the diner. She's helped herself to coffee, sits at the far end of the counter in her mountain of rags, gumming the rim of the mug. So Marietta and I are still as good as alone.

She's already behind the counter, mopping up under a cake stand, nothing in her face. I straddle the nearest stool, scared she'll move away. "I'm sorry," I say. I've never been sorrier in my life.

"I am too," she says. And I know she means she's sorry for me.

"I don't even believe it."

"What?"

"What I said."

"I don't understand," Marietta says, the rag in her hand making wet arcs on the counter, "but the worst sin in your book is somebody loving your daddy. It doesn't seem natural." She shakes her head at this—perversion. "If your mama hadn't loved your daddy, you wouldn't

[155]

even be here." She folds the wet rag in on itself.

"Your mama died last spring, didn't she?"

"Yeah, she—"

"Never mind. She died."

"That's right," Crazy Louise says down the counter. But she's not following this. She hears voices in her head and answers back.

"I never mentioned your mama," Marietta says. "Where I come from people comfort each other—with words. But I knew the first time your daddy brought you boys in here that wasn't your way. I didn't know what to make of you. Didn't know if you were too grieved to feel or too unfeeling to grieve. So I did what I could to make you all feel better about things. Every night I did my poor best—in that booth right over there.

"You think I didn't know I was playing mama for three lost souls? You used me for that. You most of all. I already knew your daddy. Passing time with his boys was only natural. Lord knows he didn't seem to know how to handle you himself. And Byron—anybody'd take pity on a little child floundering around and solemn with it. So that leaves you. And now, today, you've repaid me."

"That's right," Crazy Louise mutters. "You can't deny a thing like that."

I've got my head down now. I'm crying, mostly inside.

"You scoo—go on home now," Marietta says, as

quietly as if Crazy Louise really could understand. "I won't let on. It'll be like before. We'll finish out the summer, and there's no sense in raising sand and showing ill feeling with Byron around. You're not the only one who can use him as an excuse for doing and not doing."

I'm at the door then, half blind. I can't find the knob. My hand fights the cord on the blind. I can't get out of here quick enough, and I can't get out. "Jim?" she calls over. She's got her customer smile on; she's back in business. "Remember what I said first. I don't take it too personal. But listen here, sport"—she gives her mopping-up rag a playful little whip-crack in the air—"if I was your mama, I'd tan your bottom!"

I try to smile back. She makes a face at this pathetic effort. "And I'm old enough to be your mama too, believe it."

I'm pretty much past believing anything. I just stare at her.

"I'm thirty," she calls out, "thirteen years older than you. Where I come from that's old enough to be your mama!"

"That's right," Crazy Louise cackles. "There aren't two ways about it."

TWELVE

There are no holes to crawl into. Living through the day, of course, is not a reasonable option. It's stretching to infinity, and we're talking about eleven o'clock in the morning. I could possibly live till noon, but not with myself.

I think about rounding up Byron. Anybody who floats that well should learn a good breaststroke, a dog paddle anyhow. In a couple of weeks I could have him diving off the high board. I could . . . leave him alone.

I happen to know he's collecting lizards, his favorite dinosaur variety. I happen to know he's got a sneaker box under his bed full of lizards on lettuce leaves, like a weird salad. He studies them every night and turns them loose in the mornings and collects a new batch to study or play with or room with or whatever. A private occupation.

I know if I stand much longer at the intersection of McFarlane and Grand looking wrecked, some public-spirited pedestrian is going to call the paramedics to come for me.

I head for the Barnacle Cove—not our back lot but the high rise with the liver spots and the genuine bronze aluminum window frames. Busy as my morning has been, I haven't worked through the agenda. One thing leads to another, and there are no instant replays. Going back to explain to Marietta that we live in a laid-back age when words don't really mean anything is out. Because where she comes from . . .

Saving the day is not on the agenda.

The "uniformed-doorman-standing-at-attention-under-the-canopy," who's promised in the brochure, hasn't appeared yet. Possibly he hasn't even been born yet. I walk through the full-security doors and into the lobby. It's furnished with a color drawing of how it's going to look furnished. I walk across crumbling concrete subflooring to a button labeled *Ring for Manager to View Model Apartments.*

I ring. The grille above the button speaks in Dad's voice. He's upstairs. "Would you like to see the apartment?" his robot voice asks, tinny through static.

"Yeah," I say, possibly not in my right voice.

"Third floor. Turn right out of the elevator."

I've never been inside the Barnacle Cove before. The full blast of air conditioning feels wet on my face. In a complete transfusion real air is efficiently sucked away and replaced by sheer crispness. In this vacuum I'm about to have a word with Dad.

One doorway's open. I step inside into an entrance lined in aluminum-foil wallpaper. The place is decorated to death with electric-blue carpeting, white plaster table lamps shaped like palm trees, nubby white sofas on silver ball castors, and mirrored walls to double the room size. With the silver, blue, and white geometrics on screens, it's Dr. Zhivago's ice palace as assembled by a department store.

Dad steps out of the bedroom which is his office. "Good—"

"Morning," I finish, to cover his surprise. I'm standing there in the middle of the room, electric-blue carpeting nibbling at my sneakers, my hands planted in my back pockets.

He's standing in the doorway, rumpled but receptive. "Thought I had a customer," he says, questions standing out all over him. "Interested?" he says. "Make you a deal you can't refuse—eighty-nine five, no closing, no hidden fees. Throw in health-club membership to the right buyer."

I scan the room, glad for a role to play. "Only if you decorate."

We stand there on this neutral ground, kind of smirking at each other, groping for a beginning.

"Come on into my—office," he says finally. In there folding chairs crowd around a banged-up pine office desk, the only sticks of furniture in the place built for human use. Teetering piles of brochures carry banner headlines: THE BARNACLE COVE: THE ELEGANCE THAT BECOMES YOU.

"Catchy," I say, jerking a thumb at the slogan. "Deft double meaning."

"The advertising agency wanted to put it on T-shirts," he says. His eyes scan the ceiling to mock Madison Avenue. And then he fires a random shot. "Speaking of T-shirts, what slogan would you have on yours right now?"

I don't even have to think. I'm practically pulling it on over my head as we speak. "BORN TO BOMB OUT."

He considers this.

"And you?"

He considers this too, then nods. "Nothing too witty. I think—I HOPE I CAN SWIM BECAUSE I'M OUT OF MY DEPTH."

We gravitate to chairs, opposite each other across the desk. Buyer and seller? Patient and doctor? Repulsive brat sent to the headmaster's office?

"Why is it," Dad says, "that I have this feeling you're offering me your head on a platter?"

"I think I've already cut my own throat," I say.

[161]

"At the risk of speaking in a father's voice," he says, genuinely cautious, "do you want to talk about it?"

"I don't think I've got much choice."

"First of all," he says, "does this touch on the past, the future, or certain people we're not supposed to mention?"

"No. But screw the truce."

"The Lord giveth and the Lord taketh away," he says. I have this coming. I am the Lord.

"I've—messed up with Marietta."

"Messed up?"

"I've insulted her."

He's got his fingertips pressed together again. He's staring into them. I think he is—I don't know. My head's almost between my knees, level with the desk. Sulking again—you don't kick these habits overnight.

"So at last you've found out about Marietta," he says, which brings me around.

"Found out what?"

"That she's a human being in her own right, and nobody's easy convenience."

"You're making this too easy."

"I can make it easier. I've been there," he says. "The first night I walked into the diner, I made a pass at her."

"And?"

"I've been turned down before, but she gave the experience an entirely new meaning. She has this way of looking right through you, cutting you off at the knees, and then smiling you out the door."

He's been there all right.

"It took me a solid year to get back in her good graces."

"Then what?"

He raises his eyebrows the way he does when he's stating the obvious. "By then I was in love with her." He's even willing to close the case without a cross-examination. He knows I've taken my lumps.

Here we are man-to-man: incredible confrontation. And the room feels full of women. Marietta, Lorraine, hints of Grandmother, Mom.

Lorraine. I go for her letter in my pocket, find I've lost it.

"Lorraine's letter?" asks Dad, this all-seeing wizard. "I left it for you on the counter."

"Yeah. She's had a baby. A boy."

Into the quiet that follows this, he says, "You're full of news this morning."

"She wonders how you feel about it."

"Lorraine? How could she—"

"No. Marietta." He's linking up both my news items. I see it in his face.

"How do I feel?" he asks. "I'm a grandfather—remote, but still a grandfather. And how do I feel? I don't know. It comes too soon after being a father."

"You've had a summer's worth of that," I say.

"No, I'm having about ten minutes' worth of that."

We decide to go out looking for Byron, last seen on his hands and knees in a jungle playing ringmaster for a lizard circus. We're suspending conversation on a high—an unparalleled—peak for us. We're practically capable of slapping each other on the back. We've expanded physically. We can barely get through the door. We're also not pushing our luck.

Byron's where last seen. His rump in trunks is sticking up along the path halfway between the pool and our house. It's a well-known fact that you can't go near those little gray-beige lizards. They move at the speed of light, beyond the hand of man. One of them is climbing over Byron's wrist. Four more are peering out of weeds at him, ready to do his bidding. He's cleared a circle of dirt and is trying a diet on them: lettuce, radish bits, something that looks like Granola.

He sees us with the eyes in the back of his head. "Stay behind me. I've about got this one tame."

We do as we're told. "Can we talk?" I say, not wanting to disturb Byron's balance of nature.

"Yeah. They're deaf, I think. What are you guys doing?"

"We—ah—just wanted to tell you Lorraine's had her baby," Dad says. "A boy." It's difficult to talk to the back of a lizard trainer.

"Oh," Byron says. "It takes nine months, doesn't it." He's not angling for information.

[164]

"Yes," Dad says.

"Then I guess the nine months is up," Byron says. "This one's missing part of its tail."

Dad and I exchange looks. Our shoulders lift simultaneously. "Let's have a beer," he says.

"To celebrate?" I don't say what.

"No, let's just have a beer." We execute a wide circle around Byron into Nub's jungle and head off down to the house.

I ease up on the counter top, let my legs dangle. I'm loose as a goose. I've been low. I've been high. Now I don't know where I am. Dad pops tops, hands me a can, climbs on the stool. "Jesus," he says, looking up, "you're almost a man."

"Almost," I say. "Worst word in the tongue. Almost human. Almost scored. Almost made it."

"Almost missed out completely," he adds, turning the word.

"What am I going to do about Marietta?" I say.

"You're going to leave Marietta to me."

"You'll be in good hands. You going to marry her?"

"No," he says, into his beer. "I'm not the marrying kind, remember?"

Beautiful. I'd have to admire it even against my will, even in the heat of battle. Now he's the one offering his head on a platter. To me. But we're past that. I'm pretty sure we're at the outer limits now. "About Mom," I say.

Maybe this is too much. Maybe this is beyond the

limits. Suddenly I can't remember her before she was sick. A blank appears where her face should be, just when I can look Dad in the eye. Having them both was never in the cards, but here I am dealing out a new hand.

"Go on," he says. "It's all right."

"Why did you lea—why couldn't the two of you make it?" This is ridiculous. Like this is the only divorce on record, like I've never heard of such an incredible thing before.

He's studying the can in his hand. But he's still there with me. I study the furrows in his forehead.

"I can't come up with one specific reason. Settle for a shortcut?"

I'd settle for less than that—nothing—if that's the way he wants it. I can't say all this. I've probably said too much. I'm getting verbal as hell.

"We started too early," he says, giving each word weight.

"Didn't everybody get married early when you were . . ."

"Young," Dad finishes. "But we had an extra incentive."

"Not Grandmother," I say, a first instance of using Grandmother to lighten the tone.

"No," he says, dead serious, "not your grandmother. Lorraine."

Byron would grasp this quicker than I do. But I grasp

it. They had to get married. I make a lunge to defend Mom's honor and find she's gone.

"Later," Dad's saying, "much later, it was too easy for me to walk away from a situation I could say I never really wanted. She—your mother—had someplace to go, back with your grandmother. I knew you'd all be taken care of, and that was all the excuse I needed. I know now how gutless that was. But if I had it to do over, I can't say I wouldn't do the same thing, because I don't know."

I sit there, thinking over the beginning and the end of this marriage which I don't even take personally. I slosh beer around in the can. It seems a great opportunity to keep my mouth shut.

"You're waiting to hear more, aren't you?" he says.

"No. I'm waiting for you to tell me if I had a few more years on me and some more experience I'd be able to understand it all."

"Never crossed my mind. It's not the kind of thing I'd particularly like to hear myself," he says. And then, "Reach backward, behind you, the drawer under the knives."

I lean back, grope down the counter to the second drawer. Peer over my own shoulder as I yank it out. It's full of curled snapshots. "Pick one at random."

I pick three that are stuck together, this being Florida. What do we have here? Half-familiar scenes. The first one is Lorraine in her high-school cap and gown, gripping her diploma from the Birch Wathen School. She's standing in the little Birch Wathen cloister garden, filling it up. Another is me, Prospect Park setting, the year before acne. I'm leaning on a ball bat that comes up to my elbow. Another one of me and Byron; he's preschool. We're facing the camera like a firing squad, kid-like, against a wall. It could be any wall, but I think it's Grandmother's garage. I don't even remember the picture-taking, just details. Lorraine's cap and gown, Byron's sweat shirt, which is one of my old ones. "These are all after you left."

He nods.

"Did Mom—"

"No, after the split there was nothing. Well, lawyers, and then nothing. Grace—your grandmother—kept in touch. An occasional report, couple of lines at the most. Maybe twice a year and no replies to my replies."

"But why?"

Dad shrugged. "I'm not sure I know. Except that she has a very orderly—moral sense. I may have been on her conscience, slightly. She'd never liked me. We got off on the wrong foot. She may have thought she'd made it harder for me to stay and easier for me to go than if she hadn't been looming in the background, ready to take you over in the end.

"I never questioned it. I wanted those reports and the pictures."

He's staring into the refrigerator door and rubbing his

chin with the back of his hand. Then he pulls himself together. Complete mood change. Tilts the stool back,
hooks his thumb in a belt loop. "What I would like—
Jesus, I forgot my beeper—what I would like is for you
to meet a girl and take her out. You've been in a mon-
astery all summer."

"Longer than that," I say.

"Longer than *this summer?* Damn!" Faking amazement that anything could be longer than what we've
been through.

"A girl," I say, "is no farther away than that telephone." My hand sweeps out into the room. I nearly fall
off my perch. It's possible that I'm half drunk on a half
can of beer.

"You're talking long distance," Dad says. "You're
talking New York."

"I'm talking about a single message unit on a local
call."

"Who is she, not that I doubt you."

"Her name is . . . Adele Parker. I wrote it down
on a boarding pass, complete with phone number. She's
a fox. I met her on the plane coming down."

"And you haven't called her before?"

I fiddle with the beer can. "Before, Marietta kept
my . . . fantasies fairly well fulfilled." Dad looks my
way, and we share the sheepishness.

"You say you met her on the plane? Fast worker."

"Actually Byron took care of all that. You see, he

figured her for a stowaway and took it on himself to cover for her and—"

"Stop." Dad's grinning. "I've heard enough." We lob beer cans from immense distances into a Winn-Dixie shopping bag.

THIRTEEN

I wait till I have the house alone to dial Adele Parker's number, ring her bells. Confidence ebbs: the number's familiar, but I can't place the face.

Finally the phone's answered, twice. "Hello?" followed by a click. Adele's voice, probably, but a Mystery Guest on an extension. Will the real Adele Parker please stand up? "I've got it, *Mother*," Adele says. There's not another click, though. Mother's hanging in there.

I introduce myself again; my voice cracks for the first time in three years. "Jim Atwater!" I roar, like I'm leading cheers. Because I have this trouble saying my own name. Always have; I've stopped worrying about it. "The plane down . . . my little brother . . . a cat . . ."

Enough. She remembers. She's pleasant but guarded. Maybe she wonders what took me so long. Chooses her words with care because we're playing to an audience on this call.

After some slow-moving chitchat we both say, "Would you like—" Politely I pull back. Ladies first. I don't know what I was going to ask her to do anyway. Haven't laid the groundwork.

"—to come over here?" she says.

"Tonight?" I say, rushing my fences and wanting an excuse not to eat at Marietta's—not tonight.

Yes, tonight. I hear the monotony of months in Adele's voice. It's been a long summer. "Come on over any time. I don't eat dinner."

The extension catches its breath. Definitely her mother, scandalized at Adele's telling a stranger that proper nutrition is not practiced in that household. Dead giveaway. Adele provides address and directions. Desperate for company. Why do I regard this as a put-down instead of an opportunity? Why do I spend every minute while talking to a girl asking myself questions? "Bring your trunks," Adele says.

I go on my bike at the first sign of evening. Trunks, towel, deodorant squeeze bottle for *aprés*-swim, in flight bag dangling on a handlebar. By following directions, I leave Coconut Grove behind and sail into Coral Gables, pumping hard on the approaches to humpback bridges over yacht canals—touch of old Venice. I zip down boulevards blacked out by banyans, past lime-green, tangerine, French-gray houses sprawling on lawns with little cast-iron hitching-post boys by the drives. Carriage lamps wink encouragement.

I'm past Cremona, Sistina, Paradiso—grand boulevards all—when I'm nearly totaled by a Porsche that comes from behind out of nowhere and rockets past me into the night. I wobble on and arc into the Parkers' driveway. This must be the place. The street numbers are back-lighted at the curb and appear again on Portuguese tiles up by the front door.

The whole place is blazing with electricity. There's a small spotlight under every palm tree and big fake torches flanking the ironwork gates over the front door. I find a parking place for my bike in the three-lanes-wide drive, which is completely floodlit.

And walk between two metal boxes, where the curving path begins, which seem to trigger six or eight more invisible lights pinpointed around the lawn. It's like Christmas in New Jersey.

I press the (illuminated) doorbell, but the front door behind the iron grille is already opening and a slinky figure is coming into view. I have my smile of recognition at half mast.

And flash it full in the face of a middle-aged woman. Palm-frond shadows cross her face. A hand extends to the gate lock, but doesn't turn it. "Friend of . . ."

"Adele's," I say, giving the password. She glances down at my flight bag.

"Ah, yes, Jim . . ."

[173]

"Atwater."

"I'm her mother." The hand blazing with diamonds springs the lock. She's wearing a long orange terry-cloth —thing. Her hair, familiar auburn, is waved over one eye. She has a terminal tan, beginning to checker. A very attractive crocodile.

"We have to be so careful," she says, and plants the hand on her hollow chest. "The things you hear about . . ." She turns a hand in the air. "Come in. Adele's dying to see you." She looks at me over one shoulder. "Or is that the thing to say?" I haven't a clue.

She whisks me inside, still eyeing the flight bag. I swing it a little to indicate the absence of heavy weaponry. Still, I could have a length of wire, two gags, and a knife in there. Her eyes aren't satisfied. She closes the door behind us. It's got locks all the way to the floor, New York style. "So careful," she says. And leads me on across tile and into carpet country.

The room is so big it has furniture in the middle of it, like an island in a beige sea. Beyond that, through sliding glass doors, is a walled pool the width of the house, lighted bright blue from beneath.

Adele appears between louvered doors. It's clearly Adele: that hair, the pointed nose, the silver chain that dips in the hollow of her throat. What she's wearing starts just at her armpits and hangs to the floor. A white terry-cloth thing. "I'll take it from here, Mother," she says, which is a whole lot more intimidating to me than to Mother.

[174]

I've never been fought over by two women, and it doesn't exactly happen now. But they plant me on a sofa so low I have to look out at the world between my knee-caps, and Mother's in a velvet chair on one side, Adele on the other.

"Hi," I say to Adele, minutes too late. She rolls her green-gray eyes toward her mother, who's extended one brown leg in a gold sandal out through a slit in the orange terry cloth.

"I suppose you've seen the papers," Mrs. Parker says, replanting her hand. "The body in the bag? With the hand sticking out? Down by the bridge? Above Key Largo?"

"Mother," Adele says with immense emphasis, "they caught the killer."

"Well, I *know*." Mrs. Parker's hand climbs up to her throat. One red nail divides her chin. "I *read* it."

"Then you know Jim didn't do it. They don't get out on bail that soon."

"Oh, Adele, really." Her mother sighs and covers by turning her guns on me. We have a little give-and-take while she rechecks information: Van Cortlandt Academy, Brooklyn Heights, little brother, Summer with a Father. She tries to add to the slender store of knowledge, probes for a more recognizable family name than Atwater, learns Livingston, and nods. "And your father, he's in business down here?"

"Yes." This isn't enough. She waits. "In real estate."

She smiles, tosses her head, waves a hand. "Who isn't,

[175]

down here? And your mother's in New York?"

I nod. In a manner of speaking she is, and I have no intention of going into it. I've been going into things with people all day.

"Married again?"

I look at her, bone-tired.

"Is your mother married again?" she repeats, helping me.

"No, never again." I'm so tired I'll have to take my bike back in a cab.

"Oh, I know, I know." Both hands move in the air. "I can feel for her. The things you give up in a marriage and the little you get back." Her look flits over Adele, but she's treading on dangerous ground there and seems to know it. "Better a broken home than an unhappy one," she says, leaning confidentially my way. She's telling her daughter something for the thousandth time, this time through me.

"Hungry?" Adele says, skewering me with a look. This is a signal. It triggers Mother, who leaps out of her chair.

"Of course he is," she says. "I'll just see if I can find something." She glitters away through a dark dining room, propping a kitchen door open as she disappears through it.

"She's been arranging things on plates ever since you called," Adele murmurs.

"My mother's dead," I murmur back, which is a stupid and heavy thing to lay on her, sure to make her uncom-

[176]

fortable. She looks annoyed instead. I figure I have nothing to lose with these people. "So how's the summer going?" I say.

"Eight days till flight time," she says. "I will just make it." She states each word separately. "If I don't get back to New York pretty soon, they're going to find another body in a bag down at Key Largo." This would be impressive except she's not looking at me as she speaks.

I try again. "What have you been doing with yourself?" This is a Spence mixer in isolation. There's no wall to get against.

"We drive to Bal Harbor and have lunch at Neiman's and then we shop," she says in a voice of the dead. "We drive downtown to Omni and have lunch at Jordan Marsh, and then we shop. We drive to Dadeland, and we have lunch at Burdine's, and then we shop." She rests a weary hand at her throat, over the silver chain. I've seen this gesture before.

"We'd better swim," she says, "or it'll be three in the pool."

"Do you want to go out somewhere?" I offer, but my heart isn't in it.

"That also would be a threesome," Adele says loud enough to be heard all over the house. She stands up and electrifies me briefly by unzipping an invisible zipper down her entire front. She's wearing a swimsuit underneath. "Just change anywhere," she says, starting to wave a hand at the louvered doors, but changes the wave to a pointing finger.

[177]

"Not in your room, darling!" her mother's voice sings out from the kitchen, ears on her like an Indian scout. "In the guest room!"

Adele closes her eyes. "Third door on your left."

By the time I come back, feeling naked in trunks, the living room's empty, but the cocktail table is covered with plates of meatballs, egg rolls, cookie-cutter sandwiches, and more to come because Mother's back in the kitchen.

Adele's in the pool, lighted from beneath. Her legs are wavery under water. She's wearing a bathing cap, which is a turn-off. I hook my toes on the side, and dive in, nothing show-off. She turns on her back and drifts away. When I'm in the deep end, she's in the shallow. We reverse. I miss her in passing.

"Do you know Heidi Ames?" she says from a distance.

"Who?" I look around in the pool, thinking she's introducing someone.

"Heidi. Ames. Did I ask you before?"

"No," I say. "No."

"She's from Brooklyn Heights. Or some heights."

Old Heidi isn't going to keep us afloat. I swim down into the luminous water and break the surface in mid-pool. Adele, the water sprite, is way down by the diving board, treading water. "Andrea Barth?" she's asking as the water runs out of my ears. "She's very close with the girl who lives in the same building with that friend of yours."

"Kit Klein," I offer.

[178]

"Whoever. Do you know her?"

I stand up in the water, raise my hands to heaven. "You're the only girl I know in the world!" I announce. And with this show-stopper, her mother materializes at poolside, giving me a look. She's carrying a tray with two freezer-chilled glasses on it: Fresca with lime wedges, it turns out.

She crouches gracefully and holds the tray out over the pool so we have to swim and walk toward her, to relieve her burden. "Drinks *in* the pool," she smiles, "like *Jamaica*." Gold chains swing out from her neck, dip over the water. "And something to nibble on inside when you're finished."

We're already finished; the whole evening's finished. But as Mother staggers up and disappears back through the glass wall—to stand behind the curtains?—I give it one more shot. You don't get instant rapport overnight, I reason. And think my thinking is getting fuzzy.

We're standing nipple-deep in the cool pool, holding frosted Fresca above the blue-Jell-O water. And I'm shriveled up inside my trunks. I haven't got the patience for a Spence girl tonight. It's like talking a jumper down off a ledge, and I haven't got the patience. But what's one more shot:

"You were great with my little brother. On the plane."

She's pulling a strap up and spilling part of her drink out of the other hand. "Oh, kids . . ." she says, not able to relate the concept to Spence/New York.

It takes twenty minutes to towel-dry without the sun.

[179]

Still, the water runs down out of my trunks. Adele works a deal whereby she shimmies out of her suit inside her terry-cloth thing. And then just steps out of her suit, a black puddle on the tiles. I wrap my towel around my trunks, hoping it'll absorb the rest of the water.

We get through ten minutes by standing over the cocktail table, spearing meatballs with toothpicks and eating enough to make a respectable dent. I can feel water running between my toes. Then, looking into my old conversational grab bag and finding it empty, I go back to the guest room and get dressed.

It's a great house for lurking. I'm about to reappear through the louvered doors when I hear the tail end of some mother-daughter talk. ". . . just all right," Adele's saying through clenched teeth, "barely."

"Well, are you going to give him your number in New York?" her mother's asking. "I don't see why you shouldn't."

"Or why I should."

"If you leave this sort of thing to your father . . ."

"What sort of thing?"

"Your *social* life," her mother emphasizes.

"Is that what this is?" Adele sighs. "Oh, God."

I reappear, and they step apart.

There's a minor outburst of friendliness at the front door. Mother and daughter have joined up to see me off. Mother laughs at all the locks that have to be thrown,

[180]

switched, jiggled, keyed to get the door open. She laughs all the way to the floor. I tell her thanks. "Good to see you," I tell Adele.

"See you," she says, or echoes, or something.

They walk me out to the last lock and form a group in the gateway. Mother pushes her luck and wraps an arm temporarily around Adele's waist. They're a study in orange and white terry cloth, identical in the lawn light.

I have this feeling you have when you know you're never going to see people again, the need to observe a moment of truth. For an exit line I'd like to poke a finger at Adele and say, "You've got it all wrong here. *You're* the pain in the ass; your mother's only a minor irritation."

Instead, I'm walking backwards down the path, giving them a little salute with my flight bag. And they're giving me little waves back. Then the iron gate sighs shut. They could use a moat and a drawbridge.

And I find I can get back on my own power. I pump hard at the approaches to humpback bridges, drink in the greatest thing about Florida: that you can see clouds at night. Big thunderheads stand up over the palm trees, actually white in the night.

Then I know where I've been. I've been calling on Mrs. Got-Rocks, in the checkered flesh. And Miss Got-Rocks. But Marietta need never know. Why spoil her dream?

[181]

FOURTEEN

The door's open now, and we should be able to walk straight through it into the last days, without having to count them. What can stop us now?

I lie awake nights, fingers laced behind my head, going over events. Letting myself grieve a little, over Mom, over lost time. Balance this with plus factors. Put a few items on hold. I'd like to revise the entire Adele wipe-out into a joke. But it still stings, and the laugh track isn't there yet.

Even Grandmother turns up for review, stiff and remote as ever, almost. Twitching the see-through cane she refuses to lean on. But she's taken a quarter turn, and I see another side. We'll be back there with her soon, keeping the parlor free on Monday nights for the club to clear Nixon. I have these flashes of truth: it makes a lot of sense for Grandmother to center her cause on San Clemente. Dealing at a distance is safer, more clear-cut.

And Marietta. I'd like to talk her out of her restored cheeriness. I don't go up to the diner in these last mornings. This is understood. And at dinner, she never "lets on." If anything, she's friendlier, kinder to me than before. This hurts. But when we're talking about strong codes and the value of distance, she has a thing or two in common with Grandmother.

I even help Dad try to sell the last apartment, the eighty-nine-five number with the *Star Wars* decorating. We never actually sell the thing, but we encounter a lot of exotic browsers. And we work out a game plan. I point out the Features: nineteen-cubic-foot side-by-side freezer/refrigerator with icemaker, self-cleaning oven, splash panel above stainless-steel sink, mica cabinet top in powder room, etc. Brand-name products all. Dad talks easy financing, security system, pool privileges.

After an intense encounter with a non-English-speaking Peruvian woman and a husband, maybe hers, we collapse in our office. "We could go into business," I say, mopping my brow.

"We are in business," he says.

One-liners like that, to recognize the post-truce. We make no big deal over what we have going together. We save our extravagant claims for pushing the apartment.

I lie awake at night, in no hurry now, imagining I can hear Byron's silent lizards climbing his side of the wall. We'll have to give his room a thorough search before we leave. There's something nagging at me in the night. But I can't see it, put my finger on it. It's there, but I'm not ready for it. All the big moments seem behind us, where they ought to be.

I doze, half deaf now to the jungle sounds, the sudden rain that peppers the roof and blows into the bay. Then I'm awake again, and there's light under the door. It's like that first night of the summer. I think I hear the rise and fall of Byron's breathing through the wall, but I get up to check on things anyway. Step stealthily into shorts, vaguely contemplate burglars, vaguely contemplate how disappointed they'll be.

Dad's sitting out under the ceiling light, in pajama pants, hunched over the counter. He's got his hand around a coffee mug. His mind's elsewhere.

He doesn't see me there in the doorway. His hair's flattened in the back, standing up in front, a whitish thatch in the light. He's thick around the middle and in the upper arm. Brown shoulders fading to tan down the curve of his back. His head jerks my way, and his eyes don't know me at first.

"That stuff'll keep you up nights." I nod at the mug.

"That stuff has no effect one way or the other. Want some?"

[184]

I have the feeling I'm butting in, but I don't say no to him any more when yes is just as easy. I hang around the counter, swilling coffee. His face is clenched up, and his mind has slipped off somewhere else again. So I'm free to study him.

Maybe he can't get a decent night's sleep on that cockamamy pull-out sofa. And maybe this is the way to start. "That bed as uncomfortable as it looks?"

"Nothing wrong with that bed," he says. "I'll miss it—later."

He looks up and I recognize him finally. I see myself behind the lines around the eyes and the pockets underneath. The place he's missed shaving under his nose is the place I can never get to without nicking a nostril. Fairly routine features, but familiar. "Something wrong?" I say. I'm not about to go back to bed if he's got something he'll let me hear. We don't have that much time left, and the future's—still in the future.

He gives a little shrug, but it's more like a spasm. "I think they call it mid-life crisis." He tries to give the term a flip twist, but it falls flat.

"What?"

"Mid-life crisis. They write books about it now. Books, no cures."

"What . . . what's it like?"

He's rubbing his chin again with the back of his hand, a habit I haven't developed yet. He starts slow, talking mainly to himself. "You spend the first part of your life

[185]

running after things—grades, girls, jobs, status, roles, whatever's going.

"Then you spend the next part playing out the hand that's been dealt you. They switch the rules, and you conform to that too. Then one day you catch yourself pulling back. You lose a job you think you deserve to keep. Or you get up one morning and you can't—knot your tie around another day like the one before. Or you can't go on living another five minutes with . . . people.

"You start running away. And, if you're like me, you keep running. The running becomes the main event.

"Then one day—in a room like this—you look over your shoulder to see if anything's gaining on you. And nothing's there."

I follow this, up to a point, fall behind. "Bad feeling?"

"Empty. When you let other people down, guess who you end up feeling sorry for?"

"Yourself?"

"You got it. How?"

"I don't know. When Mom died, the way she did, I thought maybe I'd let her down." This is coming to me in words for the first time. I let them roll. "I thought—back in my mind—maybe there was something I could have done or noticed. Maybe if I'd been home the night she—went out to the car . . ."

Dad looks up. "Don't do that to yourself. You shouldn't—"

"Neither should you," I say.

And still he's worried. There's something else, and we haven't come near it. It must be Marietta. I reach out for her and prop her up between us, for the last time. "You can start over," I tell him, "with Marietta."

He's shaking his head. He's been over that with himself, and he's a little impatient with me, like Byron gets. "I would if I could. She could breathe life back into a corpse. But that wouldn't work out."

"Why not?"

"Because Marietta's—in business for herself. She's come a long way from where she's been. There's something behind her Mrs. Got-Rocks daydreams. I don't mean she gives a damn for material things, but she wants to get someplace in life. If she marries anybody, it'll be a guy from a background she can understand. Some guy trading up. Going from blue collar to white. Going from driving a rig to owning a fleet. Somebody she can move up with. You must have noticed—she pities me."

I had and I hadn't. But I hate hearing him put it into words. I stand with my finger cramped into the mug handle and wonder if getting people to talk is such great therapy. But he's talking again, something about Byron.

"—hasn't he said anything to you?" Dad's squinting up at me through the light. "He wants to stay down here, with me."

No, I didn't hear that.

"I want you to know, I didn't influence him. He—"

Not much you didn't.

"—likes it down here. I could get him into a school. He's old enough to—"

He's a baby. He's confused a vacation with real life. Hell, yes, he likes it down here. He's gone native, wandering around in the jungle like Wolf Boy.

"—begin taking charge of himself. And I can spend a lot of time with him. He's worried about what you'd think about it, Jim. He—"

He ought to be worried about what I'd think about it.

"—won't stay unless you say it's okay."

Then he won't stay.

"It's too late for you and me. But I could still be something for him, something more than just making up for lost time."

And still I haven't said anything. I've actually kept my mouth shut through all this, and the problem with that is: I've heard every word.

"I can't ask you to let me be his father, when I wouldn't be yours. But he needs one. And he'll still need one after you're away at college and grown up. I'll see him through. I promise you that."

At last I can say the word. "No." I slam the mug down, but I keep my voice under control. "No."

FIFTEEN

Senior year. Very crucial, and a drag. We're talking about the college application tension, and we're talking about Spence mixers populated by all the Adeles of this world, and their younger sisters: a whole new generation far from frost-free. And we're talking about Kit Klein lumbering out of his chrysalis as the Complete Ivy League Man: good-bye to buddydom; he's reading *Gentlemen's Quarterly* and looking for a pose. And he hasn't put in a summer like mine.

We're talking about the tall old brownstones of Brooklyn Heights, empty-eyed above their high stoops, and Grandmother's cane on the stairs on symphony night. And the West Side IRT express up to school every morning, with the same six girls in lengthening skirts who jiggle their buns out of the car at 72nd Street.

We're talking about the Lower School ball team and Advanced Placement English and the Mickey Mouse senior-slump courses. And the old school tie which is splitting at the seams with the label hanging by a thread.

And we're talking about the year Byron stayed down in Florida, the first year.

He was pretty watchful around me in those last days. He took time out from his lizards to hang out, waiting for a sign. I'd look over my shoulder, and there he'd be, silent in sneakers. Brown, lanky, "growing like a weed," in Marietta's well-worn phrase. His legs are welted with mosquito bites; his hair's bleaching out white at the ends. Scabs on his elbows, Band-Aids on his knees, fingernails like a dwarf coal miner's. I about had to wrestle him to the ground to get him to put on socks and a shirt so I could take him down to the store to get him a size larger in everything. Though who knows what they wear to school in Florida.

I don't even remember the moment when I told him he could stay. It wasn't a touching scene. I just said something like he was going to have to start making a few decisions for himself, that I was worn out making them for him. And all the while his grin's getting bigger and bigger.

"Hell," I said, "I couldn't take you back to New York looking like that anyway. You look like a damn beach rat." He grunted approvingly at himself and rubbed his chin with the back of his hand.

"And when you learn to write, I want a letter every week."

"I can write," he said. "I could write before I even went to school!" Very indignant. Outraged, in fact.

"Then be sure you prove it. I need proof."

The day they take me to the airport, we swing by the diner. And Marietta, on cue, bursts out the door and across the sidewalk. She can make time in those blob shoes of hers. She is the breeze her apron tails snap in. She throws the car door open and jumps in to sit half on my lap. Locks her hands behind my neck. "You come back now, first chance you get." Her eyes are black violet. I've done the right thing is what she's really telling me. I'm not quite the snotty little smart-mouthed Yankee creep I seemed. I am redeemed within limits. "You'll get up Nawth," she says, earnest as anything, "and you'll get so hungry for grits it'll drive you half wild."

Then, sprightly to the end, she bobs forward to brush my cheek with her lips. But my arms are already around her, and I draw her in, shift my face, and I kiss her. A real kiss. "Whoooeee," she says when I let her move away, "what them New York girls have got in store!"

Then she's out of the car, bending down to look in the window. And I search her face for a sign of real sadness. She senses this, drops her head, sticks out her lower lip. A little parody of what she knows I'm looking for. Then a wink, and she's gone. Later, on the plane, I think of her. Working the counter for a road crew, queen of her little world and dreaming of a better one. And smiling difficult cases out the door.

The three of us, Dad, Byron, and I, hang around the boarding lounge, waiting for the flight to be called, the Nueva York Champagne flight. Three men, as Marietta says, equals no conversation.

Byron's scratching a leg with the flat of his other foot, and leaning against Dad for balance. We should be laying last words on each other. Or binding the break with plans for a Christmas reunion. Byron's exploring around in his shorts pocket. There may be wildlife in there somewhere. He's pretty anxious to get back to his jungle.

They call the flight, and the boarding lounge rises up in response. We shake hands, Dad and I, Byron and I. But all around us the Latin Americans are embracing, dropping their shopping bags for a final kiss, a hug, a cry of good-bye. In a setting like this who's to notice if we put our arms around each other? We have no history of hugging, but who's to notice?

Dad puts his arms out. I put my arms out. We grapple a little. Then step together for a moment, Byron leaning against us both. We bang each other on the back, make it hearty, make it quick. Then we make the break.

RICHARD PECK was born in Decatur, Illinois. He attended Exeter University in England and holds degrees from DePauw University and Southern Illinois University.

Ghosts I Have Been, Richard Peck's seventh novel, was selected as a Best Book for Young Adults by the American Library Association; it is a companion volume to *The Ghost Belonged to Me,* an ALA Notable Book. His other books include *Representing Super Doll* and *Are You in the House Alone?* which were also ALA "Best Books."

A former resident of Brooklyn Heights, New York, Mr. Peck lives in the converted stable of a vanished mansion in Englewood, New Jersey.